Lecture Notes in Computer Science 16112

Founding Editors

Gerhard Goos
Juris Hartmanis

Editorial Board Members

Elisa Bertino, *Purdue University, West Lafayette, IN, USA*
Wen Gao, *Peking University, Beijing, China*
Bernhard Steffen , *TU Dortmund University, Dortmund, Germany*
Moti Yung , *Columbia University, New York, NY, USA*

The series Lecture Notes in Computer Science (LNCS), including its subseries Lecture Notes in Artificial Intelligence (LNAI) and Lecture Notes in Bioinformatics (LNBI), has established itself as a medium for the publication of new developments in computer science and information technology research, teaching, and education.

LNCS enjoys close cooperation with the computer science R & D community, the series counts many renowned academics among its volume editors and paper authors, and collaborates with prestigious societies. Its mission is to serve this international community by providing an invaluable service, mainly focused on the publication of conference and workshop proceedings and postproceedings. LNCS commenced publication in 1973.

Won-Ki Jeong · Hyunwoo J. Kim ·
Zhongying Deng · Yiqing Shen ·
Angelica I Aviles-Rivero · Shaoting Zhang
Editors

Foundation Models for General Medical AI

Third International Workshop, MedAGI 2025
Held in Conjunction with MICCAI 2025
Daejeon, South Korea, September 27, 2025
Proceedings

Editors
Won-Ki Jeong ⓘ
Korea University
Seoul, Korea (Republic of)

Hyunwoo J. Kim ⓘ
KAIST
Daejeon, Korea (Republic of)

Zhongying Deng ⓘ
University of Cambridge
Cambridge, UK

Yiqing Shen ⓘ
Johns Hopkins University
Baltimore, MD, USA

Angelica I Aviles-Rivero ⓘ
Tsinghua University
Beijing, China

Shaoting Zhang ⓘ
Shanghai AI Laboratory
Shanghai, China

ISSN 0302-9743   ISSN 1611-3349 (electronic)
Lecture Notes in Computer Science
ISBN 978-3-032-07844-5   ISBN 978-3-032-07845-2 (eBook)
https://doi.org/10.1007/978-3-032-07845-2

© The Editor(s) (if applicable) and The Author(s), under exclusive license to Springer Nature Switzerland AG 2026

This work is subject to copyright. All rights are solely and exclusively licensed by the Publisher, whether the whole or part of the material is concerned, specifically the rights of translation, reprinting, reuse of illustrations, recitation, broadcasting, reproduction on microfilms or in any other physical way, and transmission or information storage and retrieval, electronic adaptation, computer software, or by similar or dissimilar methodology now known or hereafter developed.
The use of general descriptive names, registered names, trademarks, service marks, etc. in this publication does not imply, even in the absence of a specific statement, that such names are exempt from the relevant protective laws and regulations and therefore free for general use.
The publisher, the authors and the editors are safe to assume that the advice and information in this book are believed to be true and accurate at the date of publication. Neither the publisher nor the authors or the editors give a warranty, expressed or implied, with respect to the material contained herein or for any errors or omissions that may have been made. The publisher remains neutral with regard to jurisdictional claims in published maps and institutional affiliations.

This Springer imprint is published by the registered company Springer Nature Switzerland AG
The registered company address is: Gewerbestrasse 11, 6330 Cham, Switzerland

If disposing of this product, please recycle the paper.

Preface

The Third International Workshop on Foundation Models for Medical Artificial General Intelligence (MedAGI) was held in Daejeon, South Korea on September 27, 2025, in conjunction with the 28th International Conference on Medical Image Computing and Computer-Assisted Intervention (MICCAI 2025).

Medical image analysis has traditionally relied on AI models trained on specific datasets, which often struggle when applied to data from different medical centers. This limitation has fueled growing interest in general medical AI—systems capable of seamlessly adapting to diverse clinical scenarios, data modalities, and task formulations encountered across hospitals and institutions. Drawing inspiration from advances in computer vision and natural language processing, foundation models—such as large language and vision-language models like GPT, LLaMA, and Gemini—have emerged as promising candidates for general AI solutions. Their remarkable proficiency across a wide range of tasks stems from massive training datasets and substantial model capacities. However, translating these successes into the medical domain, and specifically into general medical AI, remains in its early stages. Building on the success of previous years, this workshop aimed to bring together insights from the current landscape of medical AI and foundation models. In this year's workshop, we aimed to foster discussions that accelerate the transition from task-specific medical AI systems toward more generalized frameworks capable of addressing diverse tasks, datasets, and domains.

This workshop received 31 submissions, including 28 full-length manuscripts and 3 extended abstracts. All the full-length manuscripts were peer-reviewed in a double-blind format. Each manuscript was reviewed by at least three reviewers who were experts in medical image analysis. The invaluable reviews from 44 reviewers contributed to the final decision: 8 papers were accepted for oral presentation and 11 for poster presentation. Both oral and poster papers are published in this Springer LNCS volume.

The review process for extended abstracts was single-blind, conducted by the workshop chairs. The extended abstracts are not published in this volume but will be archived on the workshop website.

The success of the workshop was due to the valuable work and significant efforts of all the authors, the reviewers, the invited speakers, and the attendees. We thank all of them for their contributions. We hope that this workshop will push forward the development of general medical AI for diverse tasks, datasets, and domains.

September 2025

Won-Ki Jeong
Hyunwoo J. Kim
Zhongying Deng
Yiqing Shen
Angelica I Aviles-Rivero
Shaoting Zhang

Organization

Program Committee Chairs

Won-Ki Jeong — Korea University, South Korea
Hyunwoo J. Kim — KAIST, South Korea
Zhongying Deng — University of Cambridge, UK
Yiqing Shen — Johns Hopkins University, USA
Angelica I. Aviles-Rivero — Tsinghua University, China
Shaoting Zhang — Shanghai AI Laboratory, China

Program Committee

Gustavo Andrade-Miranda — University of Brest, France
Seungjun Baek — Korea University, South Korea
Diedre Carmo — Universidade Estadual de Campinas, Brazil
Jang-Hwan Choi — Ewha Womans University, South Korea
Se Young Chun — Seoul National University, South Korea
Woojin Chung — Hankuk University of Foreign Studies, South Korea

Erfan Darzi — Boston Children's Hospital, USA
Absalom Ezugwu — North-West University, South Africa
Chenxiao Fan — Johns Hopkins University, USA
Javier Gil-Terrón — Universitat Politècnica de València, Spain
Maria Gómez Mahiques — Universitat Politècnica de València, Spain
Xueqi Guo — Siemens Healthineers, USA
Franko Hrzic — Boston Children's Hospital, USA
Chaoyan Huang — Chinese University of Hong Kong, China
Zhifan Jiang — Children's National Hospital. USA
Jin Tae Kwak — Korea University, South Korea
Sanghyeok Lee — Korea University, South Korea
Hyungyung Lee — KAIST GSAI, South Korea
Eduard Lloret Carbonell — Shanghai Jiao Tong University, China
Lichao Mou — MedAI Technology (Wuxi) Co. Ltd., China
Jinyoung Park — Korea University, South Korea
Yiqing Shen — Johns Hopkins University, USA
Kamilia Taguelmint — University of Brest, France
Chong Wang — Beihang University, China

viii   Organization

Chaoyi Wu	Shanghai Jiao Tong University, China
Hongtao Wu	Hong Kong University of Science and Technology (Guangzhou), China
Ye Wu	Nanjing University of Science and Technology, China
Yijun Yang	Hong Kong University of Science and Technology (Guangzhou), China
Jaejun Yoo	UNIST, South Korea
Lipei Zhang	University of Cambridge, UK
Qiaoyu Zheng	Shanghai Jiao Tong University, China

Additional Reviewers

Yanqi Cheng
Joonmyung Choi
Jaewon Chu
Zhongying Deng
Peiyuan Jing
Jongha Kim
Seungkyu Kim
Anjie Le
Ji Soo Lee
Chenjia Li
Chaoyu Liu
Wentao Shi
Jing Wei Tan

Sponsors

Contents

Lamps: Learning Anatomy from Multiple Perspectives
via Self-supervision in Chest Radiographs 1
   Ziyu Zhou, Haozhe Luo, Mohammad Reza Hosseinzadeh Taher,
   Jiaxuan Pang, Xiaowei Ding, Michael B. Gotway, and Jianming Liang

Segment Anything for Cell Tracking 12
   Zhu Chen, Mert Edgü, Er Jin, and Johannes Stegmaier

BioVFM-21M: Benchmarking and Scaling Self-supervised Vision
Foundation Models for Biomedical Image Analysis 23
   Jiarun Liu, Hong-Yu Zhou, Weijian Huang, Hao Yang, Dongning Song,
   Tao Tan, Yong Liang, and Shanshan Wang

From Pathology to Radiology: Evaluating the Applicability of Pathology
Foundation Models 34
   Hyun Yang, Sumin Jung, and Jin Tae Kwak

Pathology Foundation Models Are Scanner Sensitive: Benchmark
and Mitigation with Contrastive *ScanGen* Loss 44
   Gianluca Carloni, Biagio Brattoli, Seongho Keum, Jongchan Park,
   Taebum Lee, Chang Ho Ahn, and Sergio Pereira

Improved Training Sample Efficiency and Inter-device Generalizability
in Optical Coherence Tomography Fluid Segmentation via Foundation
Models 54
   Yusuke Kikuchi, Matthew McLeod, Eric Gros, Maxime Usdin,
   Ali Boushehri, Julia Cluceru, Yaniv Cohen, Yvonna Li, and Qi Yang

Taming Stable Diffusion for Computed Tomography Blind
Super-Resolution 65
   Chunlei Li, Yilei Shi, Haoxi Hu, Jingliang Hu, Xiao Xiang Zhu,
   and Lichao Mou

RadiSimCLIP: A Radiology Vision-Language Model Pretrained
on Simulated Radiologist Learning Dataset for Zero-Shot Medical Image
Understanding 76
   Minhui Tan, Qingxia Wu, Boyang Zhang, Genqiang Ren, Jianlong Nie,
   Zhong Xue, Xiaohuan Cao, and Dinggang Shen

Improving Medical Visual Instruction Tuning with Labeled Datasets 87
*Amin Dada, Amanda Butler Contreras, Constantin Seibold,
Osman Alperen Koraş, Julius Keyl, Aokun Chen, Cheng Peng,
Alexander Brehmer, Kaleb E. Smith, Jiang Bian, Yonghui Wu,
and Jens Kleesiek*

DR.SIMON: Domain-Wise Rewrite for Segment-Informed Medical
Oversight Network 98
Seohyun Lee, Suhyun Choe, Jaeha Choi, and Jin Won Lee

The Data Behind the Model: Gaps and Opportunities for Foundation
Models in Brain Imaging 109
Salah Ghamizi, Georgia Kanli, Magali Perquin, and Olivier Keunen

LGE Scar Quantification Using Foundation Models for Cardiac Disease
Classification 120
Athira J. Jacob, Puneet Sharma, and Daniel Rueckert

Beyond Broad Applications: Can Pathology Foundation Models Adapt
to Hematopathology? 130
Chaima Ben Rabah and Ahmed Serag

EndoTracker: Robustly Tracking Any Point in Endoscopic Surgical Scene 140
*Lalithkumar Seenivasan, Joanna Cheng, Jin Fang, Jin Bai,
Roger D. Soberanis-Mukul, Jan Emily Mangulabnan,
S. Swaroop Vedula, Masaru Ishii, Gregory Hager, Russell H. Taylor,
and Mathias Unberath*

Temporally-Constrained Video Reasoning Segmentation and Automated
Benchmark Construction 150
Yiqing Shen, Chenjia Li, Chenxiao Fan, and Mathias Unberath

Cross-Modal Knowledge Distillation for Chest Radiographic Diagnosis
via Embedding Expansion, Reconstruction, and Classification 159
*DongAo Ma, Jiaxuan Pang, Ziyu Zhou, Michael B. Gotway,
and Jianming Liang*

Random Direct Preference Optimization for Radiography Report
Generation 171
*Valentin Samokhin, Boris Shirokikh, Mikhail Goncharov,
Dmitriy Umerenkov, Maksim Bobrin, Ivan Oseledets, Dmitry Dylov,
and Mikhail Belyaev*

Test-Time Adaptation of Medical Vision-Language Models 181
   Fereshteh Shakeri, Ghassen Baklouti, Julio Silva-Rodríguez,
   Maxime Zanella, Houda Bahig, Jose Dolz, and Ismail Ben Ayed

MaskedCLIP: Bridging the Masked and CLIP Space for Semi-supervised
Medical Vision-Language Pre-training 192
   Lei Zhu, Jun Zhou, Rick Siow Mong Goh, and Yong Liu

Author Index 203

💡 Lamps: Learning Anatomy from Multiple Perspectives via Self-supervision in Chest Radiographs

Ziyu Zhou[1,2], Haozhe Luo[3], Mohammad Reza Hosseinzadeh Taher[2], Jiaxuan Pang[2], Xiaowei Ding[1(✉)], Michael B. Gotway[4], and Jianming Liang[2(✉)]

[1] Shanghai Jiao Tong University, Shanghai, China
{zhouziyu,dingxiaowei}@sjtu.edu.cn,zzhou187@asu.edu
[2] Arizona State University, Tempe, USA
{mhossei2,jpang12,jianming.liang}@asu.edu
[3] University of Bern, Bern, Switzerland
haozhe.luo@unibe.ch
[4] Mayo Clinic, Phoenix, USA
Gotway.Michael@mayo.edu

Abstract. Foundation models have been successful in natural language processing and computer vision because they are capable of capturing the underlying structures (foundation) of natural languages. However, in medical imaging, the key foundation lies in human anatomy, as these images directly represent the internal structures of the body, reflecting the consistency, coherence, and hierarchy of human anatomy. Yet, existing self-supervised learning (SSL) methods often overlook these perspectives, limiting their ability to effectively learn anatomical features. To overcome the limitation, we built Lamps (*l*earning *a*natomy from *m*ultiple *p*erspectives via *s*elf-supervision) pre-trained on large-scale chest radiographs by harmoniously utilizing the consistency, coherence, and hierarchy of human anatomy as the supervision signal. Extensive experiments across 10 datasets–evaluated through fine-tuning and emergent property analysis–demonstrate Lamps' superior robustness, transferability, and clinical potential when compared to 10 baseline models. By learning from multiple perspectives, Lamps presents a unique opportunity for foundation models to develop meaningful, robust representations that are aligned with the structure of human anatomy. All code and pre-trained models are publicly released (GitHub.com/JLiangLab/Lamps).

Keywords: Self-supervised pretraining · Learn from anatomy

1 Introduction

Medical images are highly consistent across scans of the same anatomical region, especially when captured using standardized imaging protocols (e.g., CT, MRI, X-ray). Unlike photographic images, where object positions and shapes can vary significantly, medical images—such as radiological scans—show repeating organs and tissues across patients. As a result, conventional self-supervised learning (SSL) methods developed for photographic images [3,6,17] may not be optimal for medical imaging, as they fail to account for the structured and predictable patterns inherent in medical data. Although several SSL methods have been developed for medical imaging [19,21,27,28], their performance in interpreting medical images remains limited compared to the success of large language models in understanding natural language [2,7]. We believe this discrepancy arises because SSL methods designed for NLP are adept at capturing the underlying structures (or foundation) of language, allowing intrinsic properties of the language to emerge naturally [15]. In contrast, existing SSL methods for medical imaging [4,16] lack the capability to appreciate the foundational structure of medical data—human anatomy. This raises the key question: *How can we effectively learn human anatomy from abundant unlabeled medical images?*

Human anatomy in medical imaging exhibits some unique properties including (1) **Consistent spatial continuity and coherence**: The relative positions of organs and tissues remain stable due to the anatomical coherence of the human body, ensuring predictable and organized spatial relationships between structures; (2) **Anatomical structure hierarchy**: Global organs or tissues can be decomposed into smaller anatomical parts (e.g., the left lung is partitioned into superior and inferior lobes). We aim to leverage these properties as intrinsic supervision signals for pretraining in medical imaging, learning anatomical continuity, coherence, and hierarchy from three perspectives. To learn continuity, we encourage the model to learn the anatomical context exterior the given structure—for example, given the heart, the model predicts the surrounding lung region. This is termed extrapolation, where the model infers relevant information outside the provided crop without interpolation. For coherence, we guide the model to learn the consistent order of local image patches, reflecting the general consistency in the relative positions of organs and tissues. Finally, we utilize decomposable anatomies to learn hierarchy through composition and decomposition. In summary, we aim to learn anatomy from the perspectives of continuity, coherence, and hierarchy. To integrate these diverse aspects into a unified foundation model, we employ a teacher-student architecture with an innovative cyclic training strategy that distills knowledge from the student model to the teacher model.

In this research, we focus on chest X-rays (CXRs) due to their cost-effectiveness and high impact in early disease detection [22], and we have collected a large-scale dataset for anatomical learning, as shown in Fig. 1-B. Our contributions are as follows: (1) A novel approach for learning anatomical structures from multiple perspectives using unlabeled CXRs, demonstrating the significant performance enhancement achievable by integrating self-supervised learning (SSL) methods. (2) A new exploration of the emergent properties of pretrained models, including the DNA-test, cross-patient anatomical correspondence, and locality of anatomical structures. (3) A new SSL pretraining method on large-scale data, showing exceptional transferability to various target tasks in medical image analysis.

2 Learn from Anatomy

Lamps learns from scratch on unlabeled images, with the objective of yielding a common visual representation that is generalizable across disease, organ and modality. As depicted in Fig. 1-A, Lamps employs a student-teacher model, trained through a series of learning perspectives with *cyclic pretraining* to accrue anatomical knowledge. Once trained, the teacher model can be used independently for fine-tuning or zero-shot evaluations. In the following, we introduce the three learning perspectives and integrate them into one training framework.

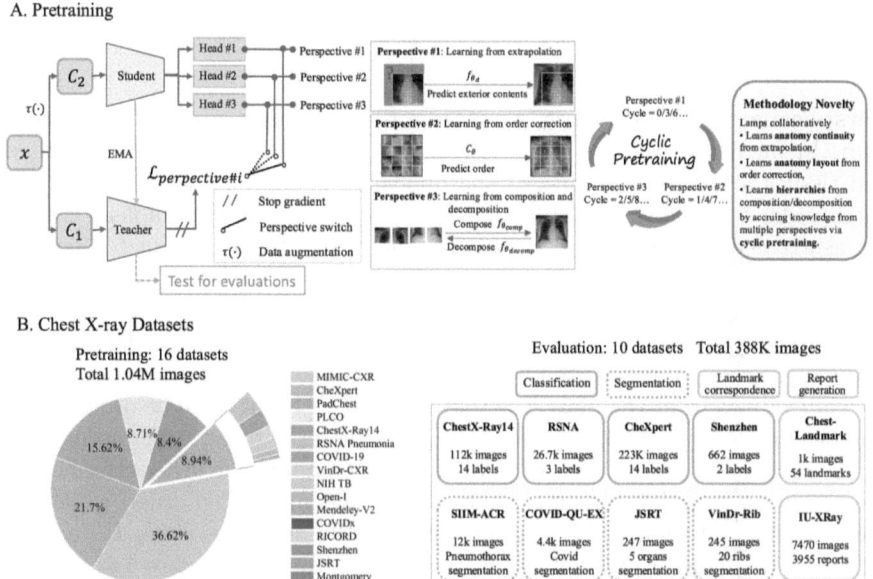

Fig. 1. We introduce Lamps, a foundation model designed to learn from multiple perspectives. A. Lamps is built using our novel framework, which accrues knowledge from three key perspectives: extrapolation, order correction, and composition/decomposition, in a cyclic pretraining process. B. Lamps is pretrained over 1.04M images and comprehensively evaluated on 10 downstream tasks.

Perspective #1: <u>Learning Anatomical Continuity via Extrapolation.</u> In this context, extrapolation refers to the model using information from a given image crop to predict semantically relevant details *outside* of that crop based on patterns learned from similar images. Our method differs significantly from interpolation and existing MIM approaches [8,24] in two key ways: (1) it does not rely on anatomical cues from unmasked patches surrounding the masked areas, resulting in more robust representations; and (2) it restores the features of the masked region rather than the original image pixels, encouraging the encoder to learn high-level representations for anatomical understanding rather than low-level superficial features for image restoration. Image pre-processing details: An input image x is divided into a grid G of 18×18 non-overlapping patches. We extract a random crop C_1 with $n_1 \times n_1$ patches from G, and an random crop C_2 with $n_2 \times n_2$ patches inside C_1, where $n_2 < n_1$, and C_1 and C_2 always begin from one of the nodes in G. C_1 is regarded as the ground truth and converted into a sequence of N_1 non-overlapping patches, where $N_1 = n_1^2$, then fed to Teacher to obtain corresponding patch-level embeddings $e_t = e_{t_1}, ..., e_{t_{N_1}}$, where e_{t_k} is the embedding associated with the k-th patch. Similarly, C_2 is input to Student and Head#1 f_{θ_d} to get the predicted embeddings $e_s = f_{\theta_d}(f_{\theta_s}(C_2, M))$, where $e_s = e_{s_1}, ..., e_{s_{N_1}}$, $M \in \{0,1\}^{n_1 \times n_1}$ denotes the set of masked patches. An l_1 loss is employed on the masked patches to align the predicted embeddings from Student with the ground truth embeddings from Teacher:

$$\mathcal{L}_{extrap} = \|e_s[M] - e_t[M]\|_1 \tag{1}$$

Perspective #2: Learning Anatomical Coherence via Order Correction. In medical imaging (e.g., chest X-ray), the relative positions of organs and tissues are consistent due to anatomical coherence. As such, the relative patch order is a strong signal for patch-level supervision. For implementation, C_1, as extracted in the last section, is randomly shuffled and input to Student, while the original C_1 (patches in order) is input to Teacher to get corresponding patch-level embeddings e_s and e_t, respectively. e_s is then passed to a prediction classifier Head#2 C_θ to get the predicted order P^o of the shuffled patches. Besides, the shuffled patched embedding from Student should be consistent with the embeddings from Teacher. The total loss of this branch will be order correction loss and local consistency loss as in Eq. 2, where T^o is the target order of the shuffled patches, $CE(a,b) = -a \log b$ is used to compute the order classification loss. e_s represents the student's patch embeddings in shuffled order, while e_t is the teacher's embeddings in the original order. We reorder e_t as $e_t[T^o]$ and use MSE loss to align the shuffled patch embeddings from Student to Teacher's.

$$\mathcal{L}_{shuffle} = \lambda * CE(T^o, P^o) + MSE(e_s, e_t[T^o]) \tag{2}$$

Perspective #3: Learning Anatomical Hierarchy via Composition and Decomposition. From a hierarchical view in anatomies, organs and tissues can be broken down into smaller components. This decomposition requires the ability to distinguish anatomical and pathological structures, allowing the model to learn the relationships between the whole anatomy and its smaller components. In detail, to learn the anatomical structure in a hierarchical way, the initial image is randomly cropped to C_1 with a cropping ratio $r = exp(-\frac{epoch}{t})$, where t increases as training progresses, enabling a coarse-to-fine crop. Inspired by [21], C_1 is divided into 4 sub-crops $C_2 = \{C_1^i\}_{i=1}^4$. To capture composition, the sub-crops are encoded by Student to get corresponding embeddings $\{e_{s_i}\}_{i=1}^4$, which are concatenated and input to a linear layer Head#3 $f_{\theta_{comp}}$ to aggregate the features: $e_{comp} = f_{\theta_{comp}}(\oplus(\{e_{s_i}\}_{i=1}^4))$, where $\oplus$ is the concatenation operator. The composition loss will be: $\mathcal{L}_{comp} = \|e_{comp} - e_t\|_1$, where e_t is the embedding of C_1 encoded by Teacher. Symmetrically, to learn decomposition, C_1 is encoded by Student to get e_s, and then input to $f_{\theta_{decomp}}$ to decompose into

4 sub-features: $e_{decomp} = f_{\theta_{decomp}}(e_s) = \{e_{decomp_i}\}_{i=1}^4$. the decomposition loss will be $\mathcal{L}_{decomp} = \frac{1}{4}\sum_{i=1}^{4}\left\|e_{decomp_i} - e_{t_i}\right\|_1$, where $\{e_{t_i}\}_{i=1}^4$ are the embeddings of $\{C_1^i\}_{i=1}^4$ encoded by Teacher. The total loss in this branch will be:

$$\mathcal{L}_{comp-decomp} = \mathcal{L}_{comp} + \mathcal{L}_{decomp} \qquad (3)$$

Training Pipeline. Training a model with a single optimization objective can only capture a limited understanding. Directly combine multiple objectives, where the total loss in each iteration is the sum of the individual losses, the model will become conflated about its optimization targets as different loss functions lead to divergent optimization directions. Therefore, we integrate the 3 perspectives into a unified framework by utilizing *cyclic pretraining* [13], which reduces computational costs compared to co-training within a single epoch and avoids the complexity of generating heterogeneous data for different tasks. For the cyclic pattern, each learning perspective will be trained once in every 3 epochs: if $epoch\%3 == 0$, $\mathcal{L} = \mathcal{L}_{extrap}$; else if $epoch\%3 == 1$, $\mathcal{L} = \mathcal{L}_{shuffle}$; else $epoch\%3 == 2$, $\mathcal{L} = \mathcal{L}_{comp-decomp}$.

3 Experiments and Results

§3.1 Experiment Protocol. Our Lamps takes the base version of the Swin transformer (Swin-B) [11] as the backbone. We adopt [3] for optimization settings (e.g., learning rate schedule, optimizer, etc.) and teacher weight updates. In Lamps, each learning branch utilizes the same student-teacher encoder architecture, with unique heads tailored to distinct learning perspectives (see Fig. 1-A). For perspective #1, $n_1 = 14, n_2 = 11$, Head#1 (f_{θ_d} is an 8-layer attention block) reconstructs the embeddings of masked patches. In perspective #2, $\lambda = 0.1$, Head#2 (C_θ is a linear layer). In the last perspective #3, Head#3 ($f_{\theta_{comp}}$ and $f_{\theta_{decomp}}$ are three-layer MLP heads). The input image size is 448^2. To demonstrate the scalability of our framework, we pretrain Lamps on both ChestX-ray14 [23] (86K images) and a large corpus of 1.04M images for 100 epochs, shown in Fig. 1-B. We compare Lamps against 10 representative publicly available methods: vision foundation models Ark [13], RAD-DINO [20], Adam-v2 [21] pretrained on 704K, 838K and 1M chest radiographs, vision-language models KAD [26], DeViDe [12] pretrained on MIMIC [10] dataset, and vision SSL models DINO [3], BYOL [6], SelfPatch [25], POPAR [18] and PEAC [27] pretrained on ChestX-ray14 [23]. We evaluate Lamps across various settings for segmentation, classification and report generation as shown in Fig. 1-B. Specifically, we investigate the *emergent* properties (localizability, anatomy correspondence and DNA-test) of Lamps using Chest-Landmark dataset (1000 test images from ChestX-ray14 dataset, manually annotated for different anatomical landmarks).

§3.2 Lamps Exhibits Emergent Properties Going Beyond Current SSL Methods. The emergent properties of foundation models are some capabilities are not explicitly programmed or anticipated during initial training. We highlight our framework's anatomy understanding capabilities by examining the unique emergent properties of Lamps' embeddings across various zero-shot settings.

*(1) **Localizability**.* Chest radiographs display consistent anatomical structures due to standardized imaging protocols and stable anatomy across individuals. We explore the foundation model's ability to convert each pixel into semantically rich embeddings, where different anatomical structures are linked to distinct embeddings, and the same structures have nearly identical embeddings across patients. To assess this, we compile the Chest-Landmark dataset with 9 anatomical landmarks (see Fig. 2-a). To assess this, we compile the Chest-Landmark dataset with 9 anatomical landmarks (see Fig. 2-a). We extract local embeddings at the landmark location, corresponding to a 32^2 resolution, from 448^2 crops centered around each landmark in the 1024^2 original images. These embeddings are visualized using a t-SNE [14] plot. In Fig. 2-a, Lamps effectively distinguishes anatomical landmarks, while other models such as Ark, RAD-DINO, and Adam-v2 produce confusing embeddings, demonstrating Lamps' ability to consistently represent semantically similar structures across patients.

*(2) **Anatomy Correspondence**.* Identifying the same anatomical structures consistently across patients helps ensure that diagnoses are based on reliable, reproducible features. To demonstrate the efficacy of our Lamps in capturing a diverse range of anatomical structures, we utilize patch-level features to query the same anatomy across different patients in zero-shot setting. In detail, we use Chest-Landmark dataset and choose $N_q = 13$ landmarks as shown in Fig. 2-b. For a given query image, we get the chosen landmarks' features $E_q = \{E_q^i\}_{i=1}^{N_q}$ the same with *Sec. (1) Localizability*. Then for the paired key image, we extract N_k patches by sliding a window of size 448^2 with a stride of 8 (zero padding for the boundary patches), then input these patches to the pretrained backbones to get a dictionary of features for the key image $E_k = \{E_k^j\}_{j=1}^{N_k}$. Finally, for each query landmark feature in E_q, we find the most similar feature in E_k with l_2 distance and the position in the key image is the predicted corresponding landmark. We compute the errors between the 13 predicted landmarks and the ground truth across the studied baselines. The errors are averaged for the 13 landmarks as shown in Table 1 and a pair of image landmark matching is visualized in Fig. 2-b. The results demonstrate that Lamps-encoded features accurately detect anatomical landmarks, capturing specific regions and maintaining consistency despite significant morphological variations.

Table 1. Anatomical structure matching error across different chest X-rays pretrained models on size of 1024^2 images.

Lamps	RAD-DINO	Adam-v2	KAD	DeViDe	POPAR	PEAC
68.17 ± 44.45	120.28 ± 150.31	94.03 ± 131.18	296.52 ± 263.65	188.21 ± 226.31	120.69 ± 134.32	95.49 ± 114.22

*(3) **DNA-test**.* We define the *DNA-test* to assess a model's ability to recognize part-whole relationships in anatomical structures, a key aspect of hierarchical

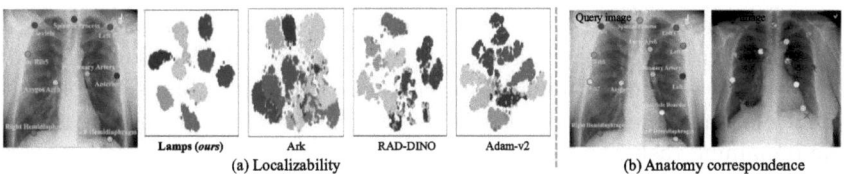

(a) Localizability   (b) Anatomy correspondence

Fig. 2. Lamps demonstrates anatomical understanding through emergent properties: (a) Localizability: distinguishing anatomical structures across patients; (b) Anatomy correspondence: accurately matching identical anatomies across patients (zero-shot predictions (red crosses) vs. ground truth (colored circles)). (Color figure online)

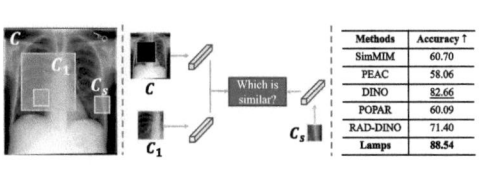
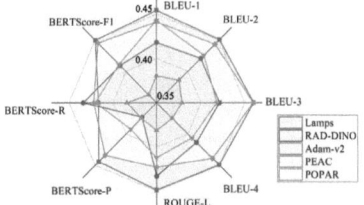

Fig. 3. Lamps learns hierarchical anatomies. Our model leverages DNA-test data to encode part-whole structures, enabling discrimination of whether a part structure belongs to its corresponding whole.

Fig. 4. Lamps outperforms baselines on 7 of 8 metrics for report generation.

understanding. For instance, it evaluates whether the model can correctly associate smaller anatomical parts (e.g., a lung lobe) with their larger structures (e.g., the entire lung). A model that can accurately discriminate part-whole relationships is likely to have a deeper and more nuanced understanding of anatomical structures. As illustrated in Fig. 3, we crop a global structure C_1 from the original image and define C as the remaining area. A smaller region, C_s, is randomly cropped from either C_1 or C. These regions are then encoded by the pretrained model (without fine-tuning) to obtain embeddings E_C, E_{C_1}, and E_{C_s}. We compute cosine similarity between E_{C_s} and both E_C and E_{C_1}, selecting the closest match to evaluate accuracy on the ChestX-ray14 test set. As shown in Fig. 3, Lamps achieves the highest accuracy, demonstrating its superior ability to capture hierarchical anatomical relationships which ensures better interpretability of the learned high-level features.

§3.3 Lamps Excels in Full Transfer Learning. We investigate the generalizability of Lamps's representations in full fine-tuning settings across a wide range of downstream tasks. To this end, in addition to training from scratch (the lower bound baseline) and a fully supervised ImageNet model, we compare two versions of Lamps—one pretrained on 86K images from the ChestX-ray14 dataset and the other on 1M chest X-ray images from 16 publicly available datasets—against 10 SOTA self-supervised and fully supervised baselines, spanning 3 organ segmentation tasks (clavicle, heart, and rib), 2 disease segmentation tasks (Pneumothorax and COVID), and 4 disease classification tasks (common thoracic diseases). As shown in Fig. 5, across all organ and disease segmentation tasks, Lamps consistently excels beyond all fully supervised and self-supervised baselines. Similarly, for classification tasks, Lamps demonstrates competitive performance in one task and significantly outperforms the baselines in the remaining tasks. In our evaluation for radiology report generation on the IU-XRay dataset [5], as shown in Fig. 4, Lamps outperforms the SSL baselines, achieving the highest scores in 7 out of 8 metrics. In summary, the consistent performance gains over baselines across various classification, segmentation, and report generation tasks underscore the versatility and generalizability of Lamps' learned representations.

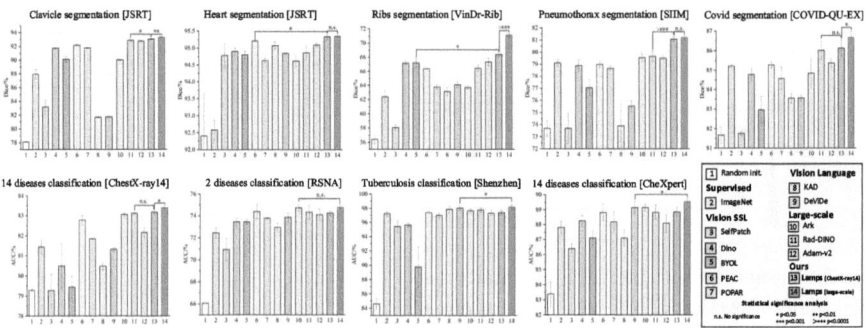

Fig. 5. Lamps achieves superior generalization and robustness, surpassing SOTA SSL methods on diverse downstream tasks. For each task, statistical significance testing ($p < 0.05$) was performed to compare Lamps with the best-performing SSL baseline.

Table 2. Ablations of individual learning perspectives and combining manners.

Learning perspective	Combination manner	Finetuning		Anatomical understanding	
		ChestX-ray14	SIIM	DNA-test	Corr-error
Extrapolation	-	82.44 ± 0.26	80.44 ± 0.32	78.70	69.33
Order correction	-	82.06 ± 0.26	79.64 ± 0.18	69.42	**65.65**
Comp-decomp	-	81.55 ± 0.21	79.73 ± 0.76	86.51	197.27
All	Directly	82.96 ± 0.24	80.77 ± 0.10	60.29	71.80
All	Cyclic training	**83.16 ± 0.05**	**81.05 ± 0.22**	**88.54**	68.17

§3.4 Ablation: Effectiveness of the Learning Objectives.

We evaluate the impact of each learning perspective in Lamps by testing individual learning perspectives and different combination methods. We fine-tune the models on the 14-class thoracic disease classification task (ChestX-ray14 [23]) and the pneumothorax segmentation task (SIIM [1]) while also evaluating them on zero-shot anatomical understanding tasks, including DNA-test and anatomy correspondence. As shown in Table 2, each branch contributes uniquely: the extrapolation branch improves fine-tuning, the order correction branch excels in landmark correspondence, and the composition-decomposition branch leads in DNA-test performance. However, simply combining the three learning perspectives by summing the losses results in degraded performance. In contrast, our cyclic training method optimally combines the branches, achieving the best performance on downstream tasks.

4 Conclusion

We introduce Lamps, a novel SSL method for visual representation learning that leverages multiple perspectives: extrapolation, order correction, composition and decomposition. These perspectives are integrated with cyclic pretraining to accrue knowledge within a Student-Teacher framework. Lamps has been rigorously tested through comprehensive experiments in various tasks, demonstrating its learned properties in comprehending anatomy and effective transferability, showing significant promise for explainable AI applications in medical image analysis.

Acknowledgments. This research has been supported in part by ASU and Mayo Clinic through a Seed Grant and an Innovation Grant, and in part by the NIH under Award Number R01HL128785. The content is solely the responsibility of the authors and does not necessarily represent the official views of the NIH. This work has utilized the GPUs provided in part by the ASU Research Computing and in part by Sol [9] and Bridges-2 at Pittsburgh Supercomputing Center through allocation BCS190015 and the Anvil at Purdue University through allocation MED220025 from the Advanced Cyberinfrastructure Coordination Ecosystem: Services & Support (ACCESS) program, which is supported by National Science Foundation grants #2138259, #2138286, #2138307, #2137603, and #2138296. The content of this paper is covered by patents pending.

References

1. sIIM-ACR pneumothorax segmentation (2019). https://www.kaggle.com/competitions/siim-acr-pneumothorax-segmentation
2. Achiam, J., et al.: GPT-4 technical report. arXiv preprint arXiv:2303.08774 (2023)
3. Caron, M., et al.: Emerging properties in self-supervised vision transformers. In: Proceedings of the IEEE/CVF International Conference on Computer Vision, pp. 9650–9660 (2021)
4. Cho, K., et al.: Chess: chest x-ray pre-trained model via self-supervised contrastive learning. J. Digit. Imaging **36**(3), 902–910 (2023)
5. Demner-Fushman, D., et al.: Preparing a collection of radiology examinations for distribution and retrieval. J. Am. Med. Inform. Assoc. **23**(2), 304–310 (2016)
6. Grill, J.B., et al.: Bootstrap your own latent-a new approach to self-supervised learning. Adv. Neural. Inf. Process. Syst. **33**, 21271–21284 (2020)
7. Guo, D., et al.: DeepSeek-R1: incentivizing reasoning capability in LLMS via reinforcement learning. arXiv preprint arXiv:2501.12948 (2025)
8. He, K., Chen, X., Xie, S., Li, Y., Dollár, P., Girshick, R.: Masked autoencoders are scalable vision learners. In: Proceedings of the IEEE/CVF Conference on Computer Vision and Pattern Recognition, pp. 16000–16009 (2022)

9. Jennewein, D.M., et al.: The sol supercomputer at arizona state university. In: Practice and Experience in Advanced Research Computing, pp. 296–301 (2023)
10. Johnson, A.E., et al.: Mimic-CXR, a de-identified publicly available database of chest radiographs with free-text reports. Sci. Data **6**(1), 317 (2019)
11. Liu, Z., et al.: SWIN transformer: Hierarchical vision transformer using shifted windows. In: Proceedings of the IEEE/CVF International Conference on Computer Vision, pp. 10012–10022 (2021)
12. Luo, H., Zhou, Z., Royer, C., Sekuboyina, A., Menze, B.: Devide: faceted medical knowledge for improved medical vision-language pre-training. arXiv preprint arXiv:2404.03618 (2024)
13. Ma, D., Pang, J., Gotway, M.B., Liang, J.: Foundation ark: accruing and reusing knowledge for superior and robust performance. In: Greenspan, H., et al. (eds.) International Conference on Medical Image Computing and Computer-Assisted Intervention, pp. 651–662. Springer (2023). https://doi.org/10.1007/978-3-031-43907-0_62
14. Van der Maaten, L., Hinton, G.: Visualizing data using T-SNE. J. Mach. Learn. Res. **9**(11), 2579–2605 (2008)
15. Manning, C.D., Clark, K., Hewitt, J., Khandelwal, U., Levy, O.: Emergent linguistic structure in artificial neural networks trained by self-supervision. Proc. Natl. Acad. Sci. **117**(48), 30046–30054 (2020)
16. MH Nguyen, D., et al.: LVM-Med: learning large-scale self-supervised vision models for medical imaging via second-order graph matching. In: Advances in Neural Information Processing Systems, vol. 36 (2024)
17. Oquab, M., et al.: Dinov2: learning robust visual features without supervision. arXiv preprint arXiv:2304.07193 (2023)
18. Pang, J., Haghighi, F., Ma, D., Islam, N.U., Hosseinzadeh Taher, M.R., Gotway, M.B., Liang, J.: POPAR: patch order prediction and appearance recovery for self-supervised medical image analysis. In: MICCAI Workshop on Domain Adaptation and Representation Transfer, pp. 77–87. Springer (2022). https://doi.org/10.1007/978-3-031-16852-9_8.
19. Pang, J., Ma, D., Zhou, Z., Gotway, M.B., Liang, J.: ASA: Learning anatomical consistency, sub-volume spatial relationships and fine-grained appearance for CT images. In: International Conference on Medical Image Computing and Computer-Assisted Intervention, pp. 91–101. Springer (2024). https://doi.org/10.1007/978-3-031-72120-5_9
20. Pérez-García, F., et al.: Rad-dino: exploring scalable medical image encoders beyond text supervision. arXiv preprint arXiv:2401.10815 (2024)
21. Taher, M.R.H., Gotway, M.B., Liang, J.: Representing part-whole hierarchies in foundation models by learning localizability composability and decomposability from anatomy via self supervision. In: Proceedings of the IEEE/CVF Conference on Computer Vision and Pattern Recognition (CVPR), pp. 11269–11281 (2024)
22. Wang, C., et al.: Prevalence and risk factors of chronic obstructive pulmonary disease in china (the china pulmonary health [cph] study): a national cross-sectional study. The Lancet **391**(10131), 1706–1717 (2018)
23. Wang, X., Peng, Y., Lu, L., Lu, Z., Bagheri, M., Summers, R.M.: CHESTX-ray8: hospital-scale chest x-ray database and benchmarks on weakly-supervised classification and localization of common thorax diseases. In: Proceedings of the IEEE Conference on Computer Vision and Pattern Recognition, pp. 2097–2106 (2017)
24. Xie, Z., et al.: SIMMIM: a simple framework for masked image modeling. In: Proceedings of the IEEE/CVF Conference on Computer Vision and Pattern Recognition, pp. 9653–9663 (2022)

25. Yun, S., Lee, H., Kim, J., Shin, J.: Patch-level representation learning for self-supervised vision transformers. In: Proceedings of the IEEE/CVF Conference on Computer Vision and Pattern Recognition, pp. 8354–8363 (2022)
26. Zhang, X., Wu, C., Zhang, Y., Xie, W., Wang, Y.: Knowledge-enhanced visual-language pre-training on chest radiology images. Nat. Commun. **14**(1), 4542 (2023)
27. Zhou, Z., Luo, H., Pang, J., Ding, X., Gotway, M., Liang, J.: Learning anatomically consistent embedding for chest radiography. In: BMVC: proceedings of the British Machine Vision Conference. British Machine Vision Conference, vol. 2023. NIH Public Access (2023)
28. Zhou, Z., et al.: Models genesis: generic autodidactic models for 3d medical image analysis. In: Shen, D., et al. (eds.) MICCAI 2019. LNCS, vol. 11767, pp. 384–393. Springer, Cham (2019). https://doi.org/10.1007/978-3-030-32251-9_42

Segment Anything for Cell Tracking

Zhu Chen(✉), Mert Edgü, Er Jin, and Johannes Stegmaier

Institute of Imaging and Computer Vision, RWTH Aachen University,
Aachen, Germany
zhu.chen@lfb.rwth-aachen.de

Abstract. Tracking cells and detecting mitotic events in time-lapse microscopy image sequences is a crucial task in biomedical research. However, it remains highly challenging due to dividing objects, low signal-to-noise ratios, indistinct boundaries, dense clusters, and the visually similar appearance of individual cells. Existing deep learning-based methods rely on manually labeled datasets for training, which is both costly and time-consuming. Moreover, their generalizability to unseen datasets remains limited due to the vast diversity of microscopy data. To overcome these limitations, we propose a **zero-shot** cell tracking framework by integrating Segment Anything 2 (SAM2), a large foundation model designed for general image and video segmentation, into the tracking pipeline. As a fully-unsupervised approach, our method does not depend on or inherit biases from any specific training dataset, allowing it to generalize across diverse microscopy datasets without fine-tuning. Our approach achieves competitive accuracy in both 2D and large-scale 3D time-lapse microscopy videos while eliminating the need for dataset-specific adaptation. The source code is publicly available at https://github.com/zhuchen96/sam4celltracking.

Keywords: Cell tracking · Foundation model · Unsupervised learning · 3D microscopy

1 Introduction

Cell tracking plays a key role in understanding tissue formation, disease progression, embryonic development, and various biomedical applications [6,21]. The task includes tracking the movement and morphological changes of individual cells, identifying parent-daughter cell pairs during mitosis, detecting cell interactions, and reconstructing full lineage information [5].

Most current methods follow the tracking-by-detection paradigm, where objects are first detected in individual frames and then linked over time to form trajectories. The linking of segmented masks is typically performed as a post-processing step, which can be solved by simple approaches like nearest-neighbor matching. Alternatively, it can be formulated as a global optimization problem and solved by integer linear programming (ILP) [14,18] or the Viterbi algorithm [17]. In this case, the performance of tracking heavily depends on the quality of the segmentation results [3]. Recently, deep learning-based methods have significantly improved both cell segmentation [12,27,30] and tracking [2,4,7–9,23]. In addition, by fine-tuning foundation models like Segment Anything (SAM) [11] and SAM2 [25] on diverse microscopy datasets, they demonstrate strong performance in 2D microscopy image segmentation [10,24]. MicroSAM [1] extends SAM for volumetric segmentation and tracking in microscopy data, but it relies on projecting 2D segmentation masks onto adjacent slices and time frames, which limits accuracy and structural analysis. Additionally, [15,31] treat different slices of 3D image data as a video sequence, allowing SAM2 to be applied for volumetric segmentation. Among these methods, training with datasets spanning multiple imaging modalities and cell types has been shown to enhance generalizability. However, microscopy image analysis faces a major challenge due to the vast diversity of cell types and imaging techniques. Methods such as phase-contrast, fluorescence, and bright-field microscopy each have unique characteristics [28]. As a result, existing methods often fail to maintain accuracy when applied to a completely different type of data.

To address the challenge, we propose a novel **zero-shot** 2D and 3D cell tracking pipeline by integrating the pretrained SAM2 model. As a fully-unsupervised approach, our method avoids dataset-specific biases and removes the requirement for manual segmentation and tracking annotations.

For 2D and small 3D datasets, we employ a linking-only approach, where we construct a complete lineage graph from a pre-segmented set of masks and correct any failed detections.

For large-scale 3D datasets, where detection and segmentation are significantly more challenging, we simultaneously address tracking and segmentation by incorporating SAM-Med3D [29], a medical foundation model trained without microscopy images. By fine-tuning it on rough segmentation masks generated by simple methods such as watershed, we enable the model to produce accurate 3D segmentation results during the tracking process. Compared to global optimization-based approaches, our tracking pipeline allows for the targeted tracking of specific subsets of cells, reducing the influence of blurring and noisy areas in large-scale images. Our key contributions are as follows:

1. With appropriate prompt selection, the pre-trained SAM2 model can be applied to segment and link cells based on previous mask information in microscopy image sequences without requiring annotated data.
2. By integrating SAM-Med3D, our pipeline achieves volumetric segmentation and tracking performance comparable to fully-supervised algorithms.

2 Method

2.1 Cell Linking in 2D and 3D Time-Lapse Microscopy Images

Cell segmentation masks are typically generated independently for each time frame and must be linked to reconstruct full lineage trees, incorporating cell migration and mitotic information. For a sequence of T raw images

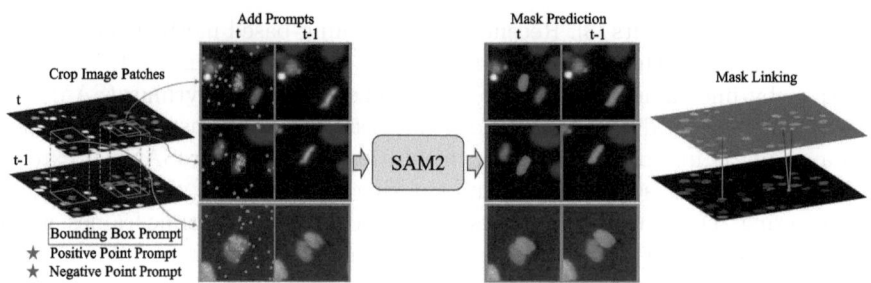

Fig. 1. Overview of the cell linking method. Patches are cropped from consecutive images based on the mask center at time point t. Bounding box and point prompts are then generated from the mask at time t. SAM2 predicts the segmentation masks of the same cell in both time frames, and the masks are linked to form tracklets. The depicted example is the Fluo-N2DL-HeLa dataset from Cell Tracking Challenge (Cell Tracking Challenge (CTC)) [19].

$I_1, I_2, \ldots, I_T \in \mathbb{R}^{w \times h}$, each image I_t has an associated segmentation mask set M_t, which can be generated using any segmentation approach. A cell lineage L_k is defined as a temporally continuous sequence of segments: $L_k = \{m_t^{k,s} \mid t_1 \leq t \leq t_2, s \in M_t\}$, where each segmentation mask $m_t^{k,s}$ of tracklet k corresponds to a specific segment s at time t. Temporal consistency ensures that each time point within $[t_1, t_2]$ contains exactly one mask per lineage L_k.

We initialize tracking at the final time frame T, where a user interface allows refinement of detections (Suppl. Video Part 1), yielding an initial mask set $\tilde{M}_T$. The algorithm then propagates masks **backward** in time, linking across frames. For each mask $m_t^{k,s}$, the center $c_t^{k,s} = (x_t^{k,s}, y_t^{k,s})$ is computed, and a square patch $p_t^{k,s}$ of side length d is extracted:

$$p_t^{k,s} = I_t \left[x_t^{k,s} - \frac{d}{2} : x_t^{k,s} + \frac{d}{2},\ y_t^{k,s} - \frac{d}{2} : y_t^{k,s} + \frac{d}{2} \right]. \tag{1}$$

The corresponding patch $p_{t-1}^{k,s}$ is extracted from I_{t-1} at the same coordinates, as depicted in Fig. 1. The pair $\{p_t^{k,s}, p_{t-1}^{k,s}\}$ is provided to SAM2 as a short video sequence. Using the known mask $m_t^{k,s}$, we generate the following prompts: 1. Bounding box $bb_t^{k,s}$ of $m_t^{k,s}$. 2. Positive points from $m_t^{k,s}$ and negative points from the background. By taking the input image patch sequence $\{p_t^{k,s}, p_{t-1}^{k,s}\}$ and applying the prompts on $p_t^{k,s}$, we apply the mask decoder to predict the potential mask for frame $t-1$. The potential mask is then compared to the pre-segmented masks in M_{t-1}. If a mask sufficiently overlaps, it is labeled as $m_{t-1}^{k,s}$ and linked to L_k. If the potential mask is too small or touches the image boundary, tracking is terminated; otherwise, if the predicted mask is valid but missing in M_{t-1}, it is added to L_k and M_{t-1} as a recovered detection. Mitosis is detected when two masks in frame t link to the same mask in $t-1$. If multiple masks link to one, only the two closest centers are retained. New tracklets are also initialized in intermediate time frames when unlinked masks remain in M_{t-1}. For 3D datasets, since the masks are volumetric, the image patches $\{p_t^{k,s}, p_{t-1}^{k,s}\}$ are extracted from the depth slice with largest mask area for each mask. Due to anisotropic voxel resolution, the x-y plane offers more reliable linking information.

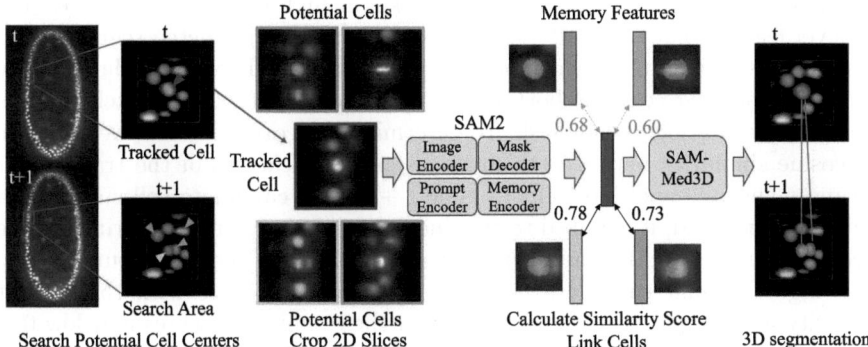

Fig. 2. For a given pair of consecutive time frames, potential cells in time point $t+1$ are searched within the neighborhood of the tracked cell in time point t. The bounding box of the tracked cell is used as a prompt for all local patches of potential cells. We use SAM2 to extract memory feature vectors for all patches, and the cell with the highest similarity score is linked. A mitotic event is detected when two potential cells have close similarity scores, then the two daughter cells are linked to the same parent cell. In the last step, 3D masks are generated by SAM-Med3D. The depicted example is the Fluo-N3DL-TRIF dataset from CTC [19].

2.2 Cell Tracking in Large-Scale 3D+t Microscopy Images

Large-scale 3D time-lapse microscopy images can reach several gigabytes per time frame and contain up to thousands of cells. Pre-segmented masks often contain missing or redundant detections, making global optimization-based linking inaccurate and inefficient for such large data. To address this, we develop a complete pipeline for simultaneous cell segmentation and tracking.

First, we apply the method introduced in [26] to detect potential cell centers for each time frame t, and generate rough segmentation masks. We then fine-tune SAM-Med3D using these rough masks, enabling it to generate 3D segmentation masks with a single click on the cell area. To facilitate refinement and selection, we provide an interactive interface (see Suppl. Video Part 1), allowing users to adjust detections or choose a specific group of cells for tracking in the initial frame.

Tracking is performed **forward** in time through each pair of consecutive time frames t and $t+1$, as illustrated in Fig. 2. For each tracked cell at time t, we search for neighboring cell centers in time frame $t+1$ within a predefined radius τ, designating them as potential candidates for linking. We then crop image patches centered on each candidate cell center and process these patches, along with the tracked cell's patch, through SAM2's image encoder to extract local visual features. Using the tracked cell's mask, we compute its bounding box, which serves as a prompt for SAM2 to generate a segmentation mask at the predicted location. The same bounding box prompt is also applied to all candidate patches to predict a potential mask of the same size as the tracked mask in the center of each patch. Then we take the high-level spatial features (the top-level feature map) from the image patches and the predicted segmentation mask and feed them to the memory encoder of SAM2. It encodes the memory representation that contains both the visual content of the local patch and the semantic information of the mask. To determine the correct match, we compute the cosine similarity between the memory-encoded features of the tracked cell at time t and all candidate cells at time $t + 1$. If a candidate cell surpasses a similarity threshold, it is linked to the lineage. In the presented experiments, we fixed this threshold to 0.8, but it might need adjustments in other circumstances. If no highly similar candidate is found, the two candidates with the highest similarity are compared. If the difference in their similarity scores is below 0.1, indicating they share local visual features but have different appearances, they are classified as daughter cells. In this case, a mitotic event is detected and two new tracklets are then initialized. If no suitable candidate is identified, the missing cell's position is predicted using the method described in Sect. 2.1. Once the matching cell center is found in time point $t+1$, the fine-tuned SAM-Med3D is applied to generate the corresponding 3D segmentation mask.

This method not only introduces new central points for missed detections but also discards redundant ones with low similarity scores. Additionally, all images are stored in Zarr format [22], allowing the whole tracking pipeline— including the fine-tuning of SAM-Med3D—to work only with small local patches rather than entire large images, significantly reducing computational costs compared to global optimization-based approaches. Furthermore, since each lineage is processed independently, the method can be parallelized, further enhancing efficiency.

3 Experiments

3.1 Metrics

As the evaluation metric for cell tracking accuracy, Acyclic Oriented Graph Matching (AOGM) [20] is used to compare predicted and ground-truth cell lineage graphs. Each tracking result is represented as a graph, where nodes represent detected cells at specific time points, and directed edges denote migration or mitosis events. AOGM quantifies tracking errors by computing the graph edit distance with weighted penalties for false positives, false negatives, missing or incorrect links, and mitosis errors. The final AOGM score is the sum of these weighted costs, where a lower score indicates higher tracking accuracy. For normalized evaluation, the tracking accuracy measure (TRA) is defined as:

$$\text{TRA} = 1 - \frac{\min(\text{AOGM}, \text{AOGM}_0)}{\text{AOGM}_0}, \qquad (2)$$

where AOGM_0 is the cost of constructing the reference graph from scratch. A TRA value close to 1 indicates better tracking accuracy. For large-scale 3D datasets, where segmentation masks are generated during tracking, we use the Jaccard-similarity-index-based segmentation accuracy measure (SEG) score to evaluate segmentation quality. It measures the Jaccard similarity index J between reference mask R and predicted mask S:

$$J(S, R) = \frac{|R \cap S|}{|R \cup S|}. \qquad (3)$$

A match is considered valid if the overlap exceeds 50% of R. The value range of SEG is between $[0, 1]$ and a higher value indicates better segmentation accuracy.

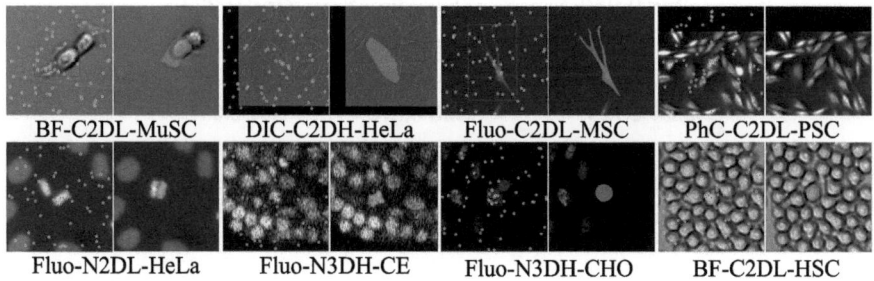

Fig. 3. Visualization of several examples of SAM2 prediction results. Left: input image patch at time t with prompts. Right: output image patch at time $t-1$ with the predicted mask. The depicted examples are from CTC [19].

3.2 Cell Linking Results

We evaluate cell linking performance using 2D and 3D datasets from the CTC [19,28]. Each dataset provides an incomplete or complete set of segmentation masks, and our goal is to reconstruct the full lineage graph. The AOGM-based linking accuracy measure (LNK) compares the predicted lineage graph with manually labeled ground truth, while the biological accuracy measure (BIO) assesses tracking performance using biologically relevant metrics such as full lineage reconstruction, longest correctly tracked segments, mitosis detection, and cell cycle duration. Since our method is based on the pretrained SAM2 model, no labeled training data for tracking is required. All parameters are selected based on the segmented cell size, without refinement with the ground truth data. Our approach is tested on 13 blind test datasets spanning different cell types and imaging techniques; the scores for selected datasets are provided in Table 1 and several examples of mask prediction are shown in Fig. 3. The hardware used in this work is an NVIDIA RTX 4090 (24 GB). We employed sam2.1_hiera_l as the pretrained SAM2 model. Tracking a sequence with 200 frames and about 100 cells per frame takes roughly one hour.

Table 1. Linking results on the selected datasets from CTC [19]; the number in the superscript represents the corresponding method; our submission is in bold.

Dataset		All 13 Datasets		Fluo-N3DH-CE		Fluo-N3DH-CHO
LNK	1st	0.984$^{(4)}$	1st	0.982$^{(4)}$	1st	0.993$^{(1)}$
	2nd	0.977$^{(1)}$	2nd	0.971$^{(1)}$	**2nd**	**0.990**
	3rd	**0.972**	**3rd**	**0.960**	3rd	0.988$^{(4)}$
BIO	1st	0.862$^{(4)}$	1st	0.862$^{(4)}$	1st	0.955$^{(1)}$
	2nd	0.794$^{(1)}$	2nd	0.782$^{(1)}$	**2nd**	**0.915**
	3rd	0.752$^{(2)}$	**3rd**	**0.667**	3rd	0.890$^{(5)}$
	5th	**0.699**				
Dataset		Fluo-N2DH-GOWT1		PhC-C2DL-PSC		Fluo-C2DL-MSC
LNK	1st	0.986$^{(4)}$	1st	0.992$^{(4)}$	1st	0.947$^{(4)}$
	2nd	**0.982**	2nd	0.991$^{(1)}$	**2nd**	**0.924**
	3rd	0.978$^{(1)}$	**3rd**	**0.991**	3rd	0.921$^{(1)}$
BIO	1st	0.788$^{(4)}$	1st	0.862$^{(4)}$	1st	0.727$^{(4)}$
	2nd	0.723$^{(2)}$	2nd	0.812$^{(1)}$	2nd	0.657$^{(1)}$
	3rd	0.718$^{(1)}$	3rd	0.805$^{(2)}$	3rd	0.647$^{(3)}$
	6th	**0.661**	**5th**	**0.792**	**6th**	**0.592**

Among the competing methods, KTH-SE (2) [16] employs a global linking algorithm and optimizes parameters using a coordinate ascent approach with ground truth data. EPFL-CH (1) [7] is a fully-supervised transformer-based network trained with tracking annotations. SIAT-CN (5) links cells using Euclidean distance and neighborhood similarity, though the exact similarity metrics were not given at the time of submission. PAST-FR (4) is an unsupervised approach based on optical flow enhanced Kalman filtering, while MON-AU (3) is also unsupervised and relies on flow network optimization. As a result, our approach achieves the top 3 in LNK score on average across all 13 datasets. Specifically, it achieves top 3 performance in 6 datasets for LNK and in 3 datasets for BIO. As a fully-unsupervised approach, our method demonstrates high performance and generalizability based on the LNK measurement, which directly evaluates linking quality. However, the BIO score is lower since our method focuses on consecutive time frames and local patches, improving efficiency but limiting the global overview of entire trajectories and cell relationships over larger areas. This can lead to breakpoints or ID switches in long trajectories, reducing long-term accuracy, which is more heavily weighted in the BIO score. Figure 4 shows several failed cases of the linking algorithm.

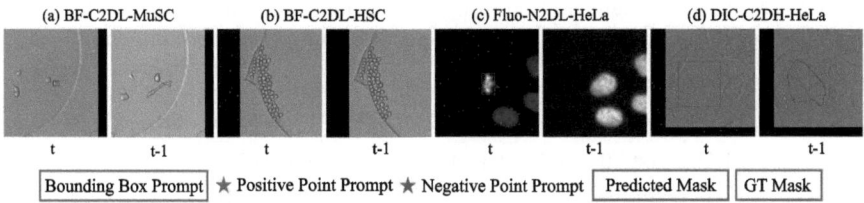

Fig. 4. Visualization of several failed cases of SAM2 prediction results. Left: input image patch at time t with prompts. Right: output image patch at time $t - 1$ with the predicted mask and the ground truth mask. The depicted examples are from CTC [19]. (a) and (c) show cases where the appearance of the cell changes significantly. In (c), the network fails to predict the mask. (b) illustrates a challenging situation with very high cell density. (d) shows that the bounding box might include neighboring cells, causing the neighboring cell to be incorrectly added.

3.3 Large-Scale 3D Cell Tracking and Segmentation Results

We evaluate our tracking and segmentation algorithm on two large-scale datasets from CTC, each containing up to around 5000 cells. Since our approach is fully-unsupervised, meaning no ground truth information from the training set is used, we apply it directly to the training set without parameter fine-tuning and evaluate its performance using ground truth annotations. For comparison, we benchmark our method against the leading approach in the current CTC rankings, as summarized in Table 2. It is important to note that this is not a direct comparison, as the competitive methods are evaluated on the challenge's official test set, while our results are based on the training set. However, based on our linking algorithm's performance, we observe that training and test set scores are highly similar in our case, since our method is unsupervised and does not overfit the training set. In the supplementary materials (Suppl. Video Part 2,3,4,5), the results for both training and test sequences are provided. It can be observed that the quality of both results is at the same level. The fine-tuning and inference were done using the same hardware mentioned previously, with SAM-Med3D-turbo employed as the pre-trained weights for segmentation.

Among the listed algorithms, KTH-SE (6) [16] employs the Viterbi algorithm for linking and utilizes ground truth data to train a multinomial logistic regression classifier for cell detection. RWTH-GE (10) [26] applies Laplacian of Gaussian-based cell detection with nearest-neighbor linking to generate the full tracklets. MPI-GE-CBG (7) is a supervised method, though its exact methodology is not publicly available. CZB-US (9) [3] uses a difference of Gaussians filter to detect cells and utilizes ILP for linking. Its hyperparameters are optimized via grid search and cross-validation on the training set. KIT-GE (8) [13] performs tracking through displacement estimation and trains a CNN-based network for segmentation, which is supervised. As a result, our approach achieves the third place in SEG score and second place in TRA score by averaging the result from both datasets, even outperforming some of the supervised methods.

Table 2. Tracking and segmentation results on the large-scale datasets from CTC [19]; the number in the superscript represents the corresponding method; our submission is in bold.

Dataset		Average Score		Fluo-N3DL-TRIC		Fluo-N3DL-TRIF
SEG	1st	$0.714^{(6)}$	1st	$0.791^{(6)}$	1st	$0.746^{(9)}$
	2nd	$0.700^{(9)}$	2nd	$0.766^{(8)}$	**2nd**	**0.695**
	3rd	**0.682**	3rd	$0.680^{(7)}$	3rd	$0.684^{(10)}$
	4th	$0.670^{(8)}$	**4th**	**0.668**	4th	$0.654^{(7)}$
TRA	1st	$0.954^{(7)}$	1st	$0.952^{(7)}$	1st	$0.955^{(7)}$
	2nd	**0.910**	2nd	$0.942^{(6)}$	2nd	$0.936^{(9)}$
	3rd	$0.908^{(6)}$	**3rd**	**0.899**	**3rd**	**0.921**
	4th	$0.895^{(9)}$	4th	$0.854^{(9)}$	4th	$0.886^{(10)}$

4 Discussion

In this work, we present a foundation-model-based **zero-shot** cell segmentation and tracking pipeline for 2D and 3D time-lapse microscopy image sequences, aiming to reconstruct full cell lineage graphs with detailed migration and mitosis analysis. We demonstrate that pre-trained foundation models can be applied to solve microscopy-specific challenges without requiring extensive fine-tuning. We incorporate SAM2 into our processing pipeline to reconstruct full tracklets from an incomplete set of pre-segmented masks, evaluating its performance across 13 different datasets. As a fully-unsupervised approach, our method achieves results comparable to state-of-the-art supervised techniques while exhibiting strong generalizability across diverse data types. Furthermore, we extend our framework to large-scale 3D+t datasets containing thousands of cells. By leveraging SAM2 and SAM-Med3D in combination with automatically generated rough detections and masks, our method efficiently segments and tracks cells while maintaining accuracy close to leading approaches.

The current approach is limited when the dataset has huge spatial movement or very large time intervals. In future work, we plan to solve these challenges by integrating motion prediction of the whole structure movement between frames. Furthermore, integrating life cycle prediction models could provide deeper insights into cell behavior, improving the biologically inspired measures of the reconstructed lineage graphs.

Acknowledgments. This project was supported by Deutsche Forschungsgemeinschaft DFG (ZC, Grant Number STE 2802/5-1).

Disclosure of Interests. The authors have no competing interests to declare that are relevant to the content of this article.

References

1. Archit, A., et al.: Segment anything for microscopy. Nat. Methods **22**, 579–591 (2025)
2. Ben-Haim, T., Raviv, T.R.: Graph neural network for cell tracking in microscopy videos. In: Avidan, S., Brostow, G., Cissé, M., Farinella, G.M., Hassner, T. (eds.) 17th European Conference on Computer Vision, ECCV, pp. 610–626. Springer (2022). https://doi.org/10.1007/978-3-031-19803-8_36
3. Bragantini, J., et al.: ULTrack: pushing the limits of cell tracking across biological scales. bioRxiv:2024.09.02.610652v1 (2024)
4. Cao, J., et al.: Establishment of a morphological atlas of the caenorhabditis elegans embryo using deep-learning-based 4D segmentation. Nat. Commun. **11**(1), 6254 (2020)
5. Friedl, P., Alexander, S.: Cancer invasion and the microenvironment: plasticity and reciprocity. Cell **147**(5), 992–1009 (2011)
6. Friedl, P., Gilmour, D.: Collective cell migration in morphogenesis, regeneration and cancer. Nat. Rev. Mol. Cell Biol. **10**(7), 445–457 (2009)
7. Gallusser, B., Weigert, M.: Trackastra: Transformer-based cell tracking for live-cell microscopy. In: Leonardis, A., Ricci, E., Roth, S., Russakovsky, O., Sattler, T., Varol, G. (eds.) 18th European Conference on Computer Vision, ECCV, pp. 467–484. Springer (2024). https://doi.org/10.1007/978-3-031-73116-7_27
8. Hayashida, J., Bise, R.: Cell tracking with deep learning for cell detection and motion estimation in low-frame-rate. In: Shen, D., et al. (eds.) MICCAI 2019. LNCS, vol. 11764, pp. 397–405. Springer, Cham (2019). https://doi.org/10.1007/978-3-030-32239-7_44
9. Hayashida, J., Nishimura, K., Bise, R.: MPM: joint representation of motion and position map for cell tracking. In: Proceedings of the IEEE/CVF Conference on Computer Vision and Pattern Recognition, pp. 3823–3832 (2020)
10. Israel, U., et al.: A foundation model for cell segmentation. bioRxiv:2023.11.17.567630v5 (2024)
11. Kirillov, A., et al.: Segment anything. In: Proceedings of the IEEE/CVF International Conference on Computer Vision, pp. 4015–4026 (2023)
12. Lee, G., Kim, S., Kim, J., Yun, S.: MEDIAR: harmony of data-centric and model-centric for multi-modality microscopy (2022). arXiv preprint arXiv:2212.03465 (2022)
13. Löffler, K., Mikut, R.: EmbedTrack–simultaneous cell segmentation and tracking through learning offsets and clustering bandwidths. IEEE Access **10**, 77147–77157 (2022)
14. Lux, F., Matula, P.: Cell tracking based on integer linear programming and probability scores. In: 2024 IEEE International Symposium on Biomedical Imaging (ISBI), pp. 1–5. IEEE (2024)
15. Ma, J., et al.: Segment anything in medical images and videos: benchmark and deployment. arXiv preprint arXiv:2408.03322 (2024)
16. Magnusson, K.E.: Segmentation and tracking of cells and particles in time-lapse microscopy. Ph.D. thesis, KTH Royal Institute of Technology (2016)
17. Magnusson, K.E., Jaldén, J., Gilbert, P.M., Blau, H.M.: Global linking of cell tracks using the VITERBI algorithm. IEEE Trans. Med. Imaging **34**(4), 911–929 (2014)
18. Malin-Mayor, C., et al.: Automated reconstruction of whole-embryo cell lineages by learning from sparse annotations. Nat. Biotechnol. **41**(1), 44–49 (2023)

19. Maška, M., et al.: The cell tracking challenge: 10 years of objective benchmarking. Nat. Methods **20**(7), 1010–1020 (2023)
20. Matula, P., Maška, M., Sorokin, D.V., Matula, P., Ortiz-de Solórzano, C., Kozubek, M.: Cell tracking accuracy measurement based on comparison of acyclic oriented graphs. PLoS ONE **10**(12), e0144959 (2015)
21. May, M., Denecke, B., Schroeder, T., Götz, M., Faissner, A.: Cell tracking in vitro reveals that the extracellular matrix glycoprotein tenascin-c modulates cell cycle length and differentiation in neural stem/progenitor cells of the developing mouse spinal cord. Biol. Open **7**(7), bio027730 (2018)
22. Miles, A., et al.: ZARR-developers/ZARR-python: v3.0.4 (2025). https://doi.org/10.5281/zenodo.14914189
23. Moen, E., et al.: Accurate cell tracking and lineage construction in live-cell imaging experiments with deep learning. bioRxiv:803205v2 (2019)
24. Na, S., Guo, Y., Jiang, F., Ma, H., Huang, J.: Segment any cell: a SAM-based auto-prompting fine-tuning framework for nuclei segmentation. arXiv preprint arXiv:2401.13220 (2024)
25. Ravi, N., et al.: SAM 2: segment anything in images and videos. arXiv preprint arXiv:2408.00714 (2024)
26. Stegmaier, J., et al.: Fast segmentation of stained nuclei in terabyte-scale, time resolved 3D microscopy image stacks. PLoS ONE **9**(2), e90036 (2014)
27. Stringer, C., Pachitariu, M.: Cellpose3: one-click image restoration for improved cellular segmentation. Nat. Methods **22**(3), 592–599 (2025)
28. Ulman, V., et al.: An objective comparison of cell-tracking algorithms. Nat. Methods **14**(12), 1141–1152 (2017)
29. Wang, H., et al.: Sam-MED3D: towards general-purpose segmentation models for volumetric medical images. arXiv preprint arXiv:2310.15161 (2023)
30. Weigert, M., Schmidt, U.: Nuclei instance segmentation and classification in histopathology images with stardist. In: 2022 IEEE International Symposium on Biomedical Imaging Challenges (ISBIC), pp. 1–4. IEEE (2022)
31. Zhu, J., Qi, Y., Wu, J.: Medical SAM 2: segment medical images as video via segment anything model 2. CoRR (2024)

BioVFM-21M: Benchmarking and Scaling Self-supervised Vision Foundation Models for Biomedical Image Analysis

Jiarun Liu[1,2,3], Hong-Yu Zhou[4], Weijian Huang[1,2,3], Hao Yang[1,2,3], Dongning Song[1,2,3], Tao Tan[5], Yong Liang[2], and Shanshan Wang[1(✉)]

[1] Paul C. Lauterbur Research Center for Biomedical Imaging, Shenzhen Institutes of Advanced Technology, Chinese Academy of Sciences, Shenzhen, China
ss.wang@siat.ac.cn
[2] Pengcheng Laboratory, Shenzhen, China
[3] University of Chinese Academy of Sciences, Beijing, China
[4] School of Biomedical Engineering, Tsinghua University, Beijing, China
[5] Faculty of Applied Sciences, Macao Polytechnic University, Macao, China

Abstract. Scaling up model and data size have demonstrated impressive improvement over a wide range of tasks. Despite extensive studies on scaling behaviors for general-purpose tasks, medical images exhibit substantial differences from natural data. It remains unclear the key factors in developing medical vision foundation models at scale. In this paper, we explored the scaling behavior across model sizes, training algorithms, data sizes, and imaging modalities in developing scalable medical vision foundation models by self-supervised learning. To support scalable pretraining, we introduce BioVFM-21M, a large-scale biomedical image dataset encompassing a wide range of biomedical image modalities and anatomies. We observed that scaling up does provide benefits but varies across tasks. Additional analysis reveals several factors correlated with scaling benefits. Finally, we propose BioVFM, a large-scale medical vision foundation model pretrained on 21 million biomedical images, which outperforms the previous state-of-the-art foundation models across 12 medical benchmarks. Our results highlight that while scaling up is beneficial for pursuing better performance, task characteristics, data diversity, pretraining methods, and computational efficiency remain critical considerations for developing scalable medical foundation models. We will open the dataset, model, and algorithms of this study at GitHub.

Keywords: Foundation models · Large medical dataset · Benchmarking · Correlation analysis · Scaling law · Self-supervised learning

1 Introduction

Biomedical imaging is at the forefront of healthcare, playing an essential role in the diagnosis and treatment of diseases [2,23]. Foundation models (FMs) offer a more efficient alternative that allows straightforward application to a wide range of tasks with minimal labeled data [2,28]. The success of FMs is largely driven by the scalable self-supervised approaches [6,18], which enable generalizable representations from large datasets without annotations [24]. The well-known scaling law [7,11] implies that the model performance can be continuously improved with increased model size, data size, and amount of computes. However, training large foundation models is expensive [5] – in computation, energy, time, and data. Unfortunately, most of the existing scaling studies are drawn from the general domain [1,5,7,11,26], while biomedical data exhibit distinct characteristics compared to general data [13,31]. Considering the diminishing returns and substantial resource requirements [7], it is important to understand the scalability in the medical domain before further scaling up.

Nevertheless, large-scale scalability research in biomedicine requires sufficient training data [11]. Despite the superior performance, many existing medical FMs and datasets are developed with a focus on a specific modality [8,9,20,22,30,32], organ [14,33], or task [3,15]. Such a focus, by its nature, maintains the existing dataset biases [10] that are limited to specific biomedical domains and reduces the opportunities to combine more data from diverse modalities [17], thus limiting the diversity and quantity of training data. The model may struggle to learn generalizable representations across different modalities and anatomical structures [17], which limits the applicability of learned scaling behavior to a wider spectrum of medical applications.

In this work, we explore the scaling behavior for developing self-supervised medical vision FMs. Our empirical experiments reveal that scaling up did take benefits but differed across tasks, whereby optimal scaling strategy depends on task and dataset characteristics. Our contributions are summarized as follows:

- We curate BioVFM-21M, a large biomedical image dataset with over 21 million images, spanning 10 imaging modalities and 30 anatomical structures.
- We conducted extensive experiments with an additional correlation analysis in exploring the scalability across model sizes, training algorithms, data sizes, and imaging modalities in developing self-supervised medical vision FMs.
- We propose BioVFM, a self-supervised medical vision foundation model at scale, demonstrating remarkable performance across 12 medical benchmarks.

2 The BioVFM-21M dataset

To support the training of BioVFM, we curated BioVFM-21M, a large biomedical image dataset consisting of 21 million biomedical images from 61 publicly available data sources. BioVFM-21M following three key principles: *1) large-scale*: As shown in Fig. 1b, BioVFM-21M is composed of 21 million images, which is significantly larger than previous datasets sourced from medical image datasets or medical websites. *2) data diversity*: As illustrated in Fig. 1a and 1c, BioVFM-21M cover most of the major anatomical regions of the human body from microcosmic to macroscopic across 10 biomedical imaging modalities, including CT, MR, Microscopy, X-Ray, Dermoscopy, Ultrasound, Endoscopy, Fundus, Nuclear

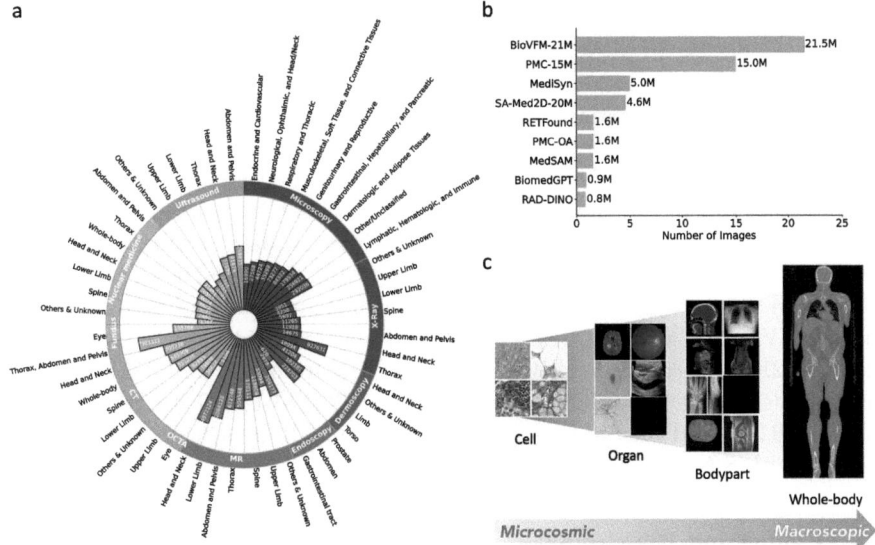

Fig. 1. The proposed BioVFM-21M dataset. a. BioVFM-21M covers a wide range of imaging modalities and anatomical areas. The illustration is exponential-scaled. **b.** Comparison with existing large-scale biomedical image datasets. **c.** BioVFM-21M covers images from microcosmic to macroscopic.

medicine, and OCTA. *3) publicly available*: We collected data from 61 publicly available sources to ensure transparency and accessibility.

To reduce the risk of data leakage, we excluded 1.8 million testing and validation images based on default data splits of each source dataset. In addition, all data sources related to our evaluation benchmarks were excluded from BioVFM-21M, ensuring that performance is not inflated by overlapped training and evaluation data. For 3D images, we slice them along axial, sagittal, and coronal views. We performed a human evaluation by randomly reviewing images from each data source. The initial version of our dataset has over 30 million images, while all of the detected padding slices were removed since it does not have anatomical structure information. Furthermore, we removed images with aspect ratio smaller than 0.4. All images were normalized to 0–255 and resized to 224 × 224px while preserving their original aspect ratio.

3 Method

3.1 Self-supervised Pretraining

We investigate the scalability of the self-supervised medical vision foundation models by varying model sizes, training algorithms, data numbers, and imaging modalities. Specifically, we select two representative self-supervised learning methods for model and algorithm scalability experiments with the ViT architecture: MAE [6] (BioVFM-M) and DINO V2 [18] (BioVFM-D). These methods are trained with model sizes ranging from 5 million to 303 million parameters while other settings follow the default settings as the author suggests. To optimize computational resources, we pretrain these models on a subset of 2.45 million images drawn from diverse imaging modalities and anatomical structures, sourced from [12,16]. In the data and modality scaling experiments, we extract two smaller subsets from BioVFM-21M with different data numbers and image modalities, which are BioVFM-US and BioVFM-0.2M. The BioVFM-US is an ultrasound subset with 0.4 million ultrasound images from BioVFM-21M. Similarly, BioVFM-0.2M is another small subset that was randomly sampled from BioVFM-21M but contains all 10 image modalities from BioVFM-21M. We train three models with 303 million parameters using BioVFM-D on BioVFM-US, BioVFM-0.2M, and the full BioVFM-21M dataset. All models are pretrained on a server with 8 NVIDIA A100 80GB GPUs.

3.2 Evaluation Benchmarks

In contrast to previous scaling studies that evaluate loss values, we assess model performance using the Area Under the Curve (AUC) score across 12 diagnostic benchmarks from MedMNIST [25]. The 12 benchmarks span a variety of imaging modalities, anatomical regions, number of classes (ranging from 2 to 14), sample numbers (from 100 to 100,000), and task types (binary/multi-class classification, multi-label classification). For evaluation, we finetune the pretrained models using linear-probing settings with an NVIDIA 2080Ti GPU by default. We specify the learning rate by 0.01 with AdamW optimizer, a batch size of 128, and image size of 224 × 224px. The maximum epoch number is 100 and early stopping is applied based on validation loss. These hyperparameters were chosen based on preliminary experiments, which demonstrated optimal performance. In addition, we compare the performance of BioVFM with ImageNet pretrained ViTs (IN21K-ViT-Base, IN21K-ViT-Large), RAD-DINO [20], BiomedCLIP [29], and BiomedGPT [29], which differ in pretraining data size (0.9 to 21 million), supervision (self-supervised learning (SSL), supervised learning (SL), vision-language pretraining (VLP)), model size (85.8 to 303 million), and data sources (natural or medical data).

3.3 Scalability Analysis

To quantify the benefits of model scaling, we fit a power function $y = e^b x^a$ as proposed in [21]. Here a indicates the slope of the scaling curve and b is the corresponding intersection. A higher slope a indicates that scaling up the model size has a more positive effect on performance for the corresponding task. Additionally, to investigate the underlying factors related to scalability, we extract 4 quantitative metrics from downstream tasks: the number of samples, Davies-Bouldin Index (DBI) [4], the number of classes, and g-zip compressibility [19]. The number of samples and the number of classes provide fundamental insights

Table 1. Average performance on 12 MedMNIST benchmarks. The values in brackets indicate 95% CI using the nonparametric bootstrap method with 1,000 resamples. †: Developed with classification labels. ‡: Developed with image-text pairs. ∗: Developed for a specific medical domain. MCC: Matthews Correlation Coefficient. BA: Balanced Accuracy.

Model	MCC	BA	F1	AUC
IN21K-ViT-Base[†]	61.59 (59.98,62.97)	69.46 (68.53,70.35)	67.16 (66.39,67.86)	91.86 (91.18,92.43)
IN21K-ViT-Large[†]	63.33 (61.78,64.73)	70.87 (69.83,71.80)	67.79 (66.97,68.54)	92.26 (91.52,92.85)
RAD-DINO [20]∗	62.88 (61.38,64.26)	69.43 (68.50,70.32)	67.07 (66.33,67.77)	91.49 (90.65,92.21)
BiomedGPT [27]‡	64.71 (63.25,66.04)	71.92 (71.01,72.81)	68.63 (67.85,69.42)	92.86 (92.23,93.46)
BiomedCLIP [29]‡	66.60 (65.19,67.86)	73.24 (72.36,74.15)	71.29 (70.56,72.01)	93.52 (92.95,94.00)
BioVFM	**69.92** (**68.52,71.12**)	**76.05** (**75.09,76.92**)	**73.43** (**72.55,74.25**)	**94.40** (**93.90,94.84**)

into the dataset's characteristics, while DBI reflects the complexity of data distribution and classification difficulty. The g-zip compressibility, on the other hand, quantifies the information redundancy of the data. With these metrics, we analyze the correlation between the computed values and the scaling slopes a using the Pearson Correlation Coefficient (PCC) in Sect. 4.5.

4 Results

4.1 BioVFM: Biomedical Vision Foundation Model at Scale

Based on the BioVFM-21M dataset, we propose BioVFM, a Biomedical Vision Foundation Model at scale. By pretraining on 21 million biomedical images with 630 million parameters using BioVFM-D, BioVFM sets a new benchmark for generalizable medical foundation models. As shown in Table 1, BioVFM significantly outperforms existing medical foundation model BiomedGPT [27], BiomedCLIP [29], and RAD-DINO [20] across 12 medical benchmarks with linear-probing by at least 3.32% in MCC, 2.81% in BA, 2.14% in F1 score, and 0.94% in AUC. Our results highlight the advantages of scalable self-supervised pretraining in the development of generalizable medical foundation models.

4.2 Model Scaling

As shown in Fig. 2a and Fig. 3, when scaling up the model size from 5 to 303 million, both BioVFM-D and BioVFM-M demonstrate improved mean AUC as the scaling law predicted. The average performance of BioVFM-D was improved by

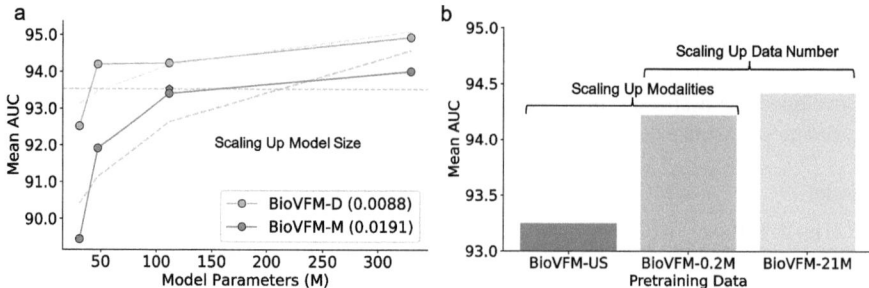

Fig. 2. Medical FMs scales with model size, data size, and number of modalities. a. Model scaling of BioVFM-D and BioVFM-M. The dotted line indicates the fitted scaling curves and the value within brackets indicates the slope a. The turquoise star indicates the performance of BiomedCLIP. **b.** Data and modality scaling with BioVFM-D. The results are averaged over 12 medical benchmarks.

2.1% and BioVFM-M was improved by 2.3% through scaling up model size. However, scaling up model size does not guarantee consistent improvements across all medical tasks. As shown in Fig. 4, the scaling slope across different tasks and imaging modalities shows considerable variation, with differences exceeding 40-fold. For example, increasing model size from 5.5 to 303.3 million improves AUC by 4.65% on ChestMNIST, whereas performance plateaus on RetinaMNIST when model sizes beyond 21.7 million parameters.

Interestingly, we observe a similar trend in the relative scaling slopes across different evaluation benchmarks. Specifically, for each benchmark, the computed scaling slopes may vary, but the relative order of the slopes remains stable: the top 5 benchmarks with higher slopes consistently exhibit greater scaling benefits compared to those with lower slopes, regardless of the pretraining method used. The gap between high and low slope groups ranges from 3 to 14 folds, highlighting significant variability in scaling efficiency. However, estimating scaling benefits based on qualitative factors, such as imaging modality, anatomy, or view orientation, remains challenging. For instance, although ChestMNIST and PneumoniaMNIST share the same imaging modality and anatomical region, they show different scaling slopes. OrganAMNIST, OrganCMNIST, and OrganSMNIST are derived from the same CT data but in different views, they demonstrate similar scaling slopes across all three tasks. These findings suggest that more nuanced factors, such as dataset size, task complexity, and intra-class variability, are likely to influence the scalability of vision foundation models.

4.3 Data and Modality Scaling

As shown in Fig. 2b, scaling up image modalities leads to a noticeable improvement of 0.97% in the mean AUC across 12 medical diagnosis tasks, demonstrating the importance of data diversity for enhancing generalizability. Surprisingly, we found that improving performance by scaling imaging modalities does

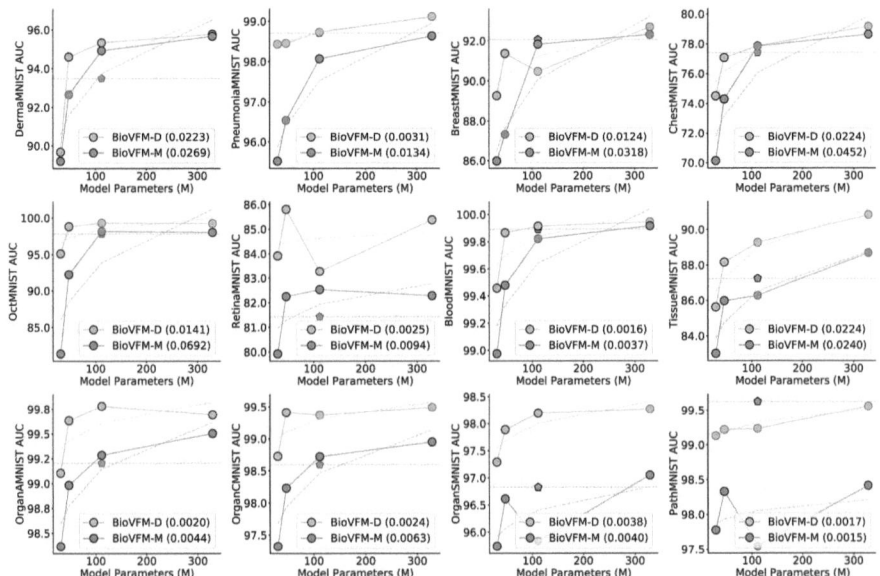

Fig. 3. Model scaling results across 12 medical benchmarks. The dotted line indicates the fitted scaling curves and the value within brackets indicates the slope a. The turquoise star indicates the performance of BiomedCLIP.

not necessarily degrade the performance of specific tasks. The model trained on BioVFM-0.2M outperforms the model trained on BioVFM-US on the ultrasound benchmark (92.75% vs. 91.87% in AUC). Furthermore, simply scaling up the data size can yield diminishing returns, with only a 0.2% improvement in mean AUC despite a 100-fold increase in data size. These findings suggest that data diversity, rather than just data size, plays a more significant role in developing scalable and generalizable medical foundation models.

4.4 Scalability Vary Across Pretraining Algorithms

We compare the self-supervised pretraining algorithms by two dimensions: the scaling efficiency (quantified by scaling slope) and the performance on medical benchmarks. As shown in Fig. 2a, BioVFM-M exhibits a steeper scaling slope (0.0191) compared to BioVFM-D (0.0088), suggesting it benefits more from increased model capacity. However, as shown in Fig. 3, BioVFM-D outperforms BioVFM-M in almost all the benchmarks. The average gap between BioVFM-D and BioVFM-M across all benchmarks and model sizes is 1.7%. Notably, BioVFM-D outperforms BiomedCLIP on 7 out of 12 tasks even when using 3.5× fewer parameters. This divergence underscores the critical role of pertaining objective design in determining both scalability and downstream task performance when developing medical foundation models.

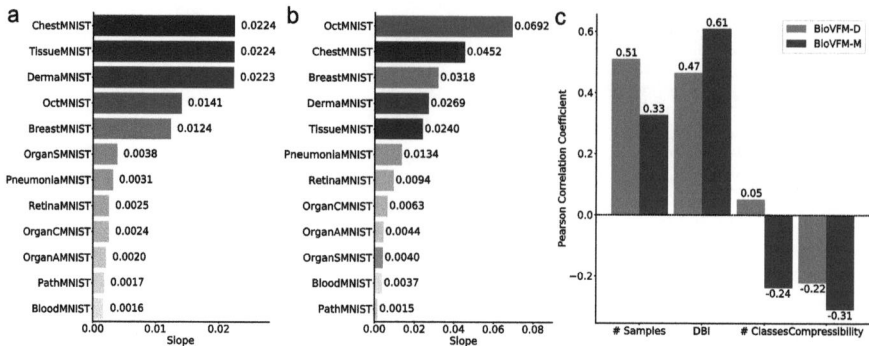

Fig. 4. Model scaling slopes and correlation to benchmark properties. **a.** The model scaling slopes of BioVFM-D with 12 medical benchmarks. **b.** The model scaling slopes of BioVFM-M with 12 medical benchmarks. The colors are consistent for each task. **c.** The Pearson correlation coefficient between model scaling slopes and benchmark metrics.

4.5 Correlations Between Benchmark Properties and Scaling Slopes

To investigate the underlying impact factors related to scaling efficiency, we take model scaling as an example and perform a Pearson correlation coefficient analysis between quantitative metrics and scaling slopes. As shown in Fig. 4c, the number of samples and DBI demonstrate a stronger correlation with model scaling slopes. A higher DBI indicates that the data has less distinct clusters, which increases the classification difficulty. This complexity necessitates larger models to capture complex decision boundaries, thus demonstrating better scaling efficiency. Additionally, the number of samples is positively correlated with model scaling slopes, as larger models can remember more information from more data to improve performance. Another meaningful indicator is g-zip compressibility, which measures the ratio of the compressed data size to the original size in bytes. Images with more redundancy will exhibit lower g-zip compressibility scores. Interestingly, we observed a negative correlation between g-zip compressibility and model scaling slopes, suggesting that larger models are more effective in handling image redundancy. In summary, larger models are advantageous for complex tasks, such as intricate decision boundaries and high data redundancy. However, simpler tasks should balance the diminishing returns of scaling with the associated computational overhead.

5 Discussion

This paper presents a systematic empirical study on the scalability of self-supervised learning for medical vision foundation models. We develop a large-scale medical image dataset to support this study. We investigate the scaling behavior over model sizes, data sizes, image modalities, and pretraining algorithms. Our results reveal that while scaling up did demonstrate benefits for medical foundation models, several important factors including pretraining algorithm, data diversity, or task complexity contribute to the varied scaling behaviors. We believe that this study paves the way for future medical AI systems that can adapt to a wide range of clinical tasks and settings. Finally, we propose BioVFM, a medical vision foundation model at scale, demonstrating superior performance across 12 medical benchmarks. We hope this study provides practical insights for building scalable and generalizable medical foundation models.

Acknowledgments. This research was partly supported by the National Natural Science Foundation of China (No. 62222118, No. U22A2040), Shenzhen Medical Research Fund (No. B2402047), the major key project of Pengcheng Laboratory under grant PCL2023AS1-2, PCL2024A06-3, PCL2024AS103, and 040202523002, National Key R&D Program of China (No. 2023YFA1011400), Key Laboratory for Magnetic Resonance and Multimodality Imaging of Guangdong Province (No. 2023B1212060052), and Youth Innovation Promotion Association CAS.

Disclosure of Interests. The authors have no competing interests to declare that are relevant to the content of this article.

References

1. Alabdulmohsin, I.M., Zhai, X., Kolesnikov, A., Beyer, L.: Getting ViT in shape: Scaling laws for compute-optimal model design. In: Advances in Neural Information Processing Systems, vol. 36, pp. 16406–16425. Curran Associates, Inc. (2023)
2. Azad, B., et al.: Foundational models in medical imaging: a comprehensive survey and future vision. arXiv preprint arXiv:2310.18689 (2023)
3. Butoi, V.I., Ortiz, J.J.G., Ma, T., Sabuncu, M.R., Guttag, J., Dalca, A.V.: UniverSeg: universal medical image segmentation. In: Proceedings of the IEEE/CVF International Conference on Computer Vision, pp. 21438–21451 (2023)
4. Davies, D.L., Bouldin, D.W.: A cluster separation measure. IEEE Trans. Pattern Anal. Mach. Intell. **PAMI-1**(2), 224–227 (1979). https://doi.org/10.1109/TPAMI.1979.4766909
5. Hägele, A., Bakouch, E., Kosson, A., Allal, L.B., Von Werra, L., Jaggi, M.: Scaling laws and compute-optimal training beyond fixed training durations. arXiv preprint arXiv:2405.18392 (2024)
6. He, K., Chen, X., Xie, S., Li, Y., Dollár, P., Girshick, R.: Masked autoencoders are scalable vision learners. In: 2022 IEEE/CVF Conference on Computer Vision and Pattern Recognition (CVPR), pp. 15979–15988 (2022)
7. Hoffmann, J., et al.: Training compute-optimal large language models. In: Proceedings of the 36th International Conference on Neural Information Processing Systems. NIPS 2022, Curran Associates Inc., Red Hook, NY, USA (2022)
8. Huang, W., et al.: Enhancing representation in radiography-reports foundation model: a granular alignment algorithm using masked contrastive learning. Nat. Commun. **15**(1), 7620 (2024). https://doi.org/10.1038/s41467-024-51749-0
9. Huang, W., et al.: Enhancing the vision-language foundation model with key semantic knowledge-emphasized report refinement. Med. Image Anal. **97**, 103299 (2024). https://doi.org/10.1016/j.media.2024.103299

10. Jones, C., Castro, D.C., Sousa Ribeiro, F., Oktay, O., McCradden, M., Glocker, B.: A causal perspective on dataset bias in machine learning for medical imaging. Nat. Mach. Intell. **6**(2), 138–146 (2024). https://doi.org/10.1038/s42256-024-00797-8
11. Kaplan, J., et al.: Scaling laws for neural language models. arXiv preprint arXiv:2001.08361 (2020)
12. Lin, W., et al.: PMC-CLIP: contrastive language-image pre-training using biomedical documents. In: Medical Image Computing and Computer Assisted Intervention – MICCAI 2023, vol. 14227, pp. 525–536. Springer Nature Switzerland (2023). https://doi.org/10.1007/978-3-031-43993-3_51
13. Liu, J., et al.: Swin-UMamba: adapting mamba-based vision foundation models for medical image segmentation. IEEE Trans. Med. Imag., 1 (2024). https://doi.org/10.1109/TMI.2024.3508698
14. Liu, P., Puonti, O., Hu, X., Alexander, D.C., Iglesias, J.E.: Brain-ID: learning contrast-agnostic anatomical representations for brain imaging. In: Leonardis, A., Ricci, E., Roth, S., Russakovsky, O., Sattler, T., Varol, G. (eds.) European Conference on Computer Vision, pp. 322–340. Springer (2024). https://doi.org/10.1007/978-3-031-73254-6_19
15. Ma, J., He, Y., Li, F., Han, L., You, C., Wang, B.: Segment anything in medical images. Nat. Commun. **15**(1), 654 (2024). https://doi.org/10.1038/s41467-024-44824-z
16. Mei, X., et al.: RadImageNet: an open radiologic deep learning research dataset for effective transfer learning. Radiol. Artif. Intell. **4**(5) (2022). https://doi.org/10.1148/ryai.210315
17. Moor, M., et al.: Foundation models for generalist medical artificial intelligence. Nature **616**(7956), 259–265 (2023). https://doi.org/10.1038/s41586-023-05881-4
18. Oquab, M., et al.: DINOv2: Learning robust visual features without supervision. Trans. Mach. Learn. Res. **2024** (2024). https://openreview.net/forum?id=GLm1BA3C8p
19. Pandey, R.: GZIP predicts data-dependent scaling laws. arXiv preprint arXiv:2405.16684 (2024)
20. Pérez-García, et al.: Exploring scalable medical image encoders beyond text supervision. Nat. Mach. Intell., 1–12 (2025). https://doi.org/10.1038/s42256-024-00965-w
21. Porian, T., Wortsman, M., Jitsev, J., Schmidt, L., Carmon, Y.: Resolving discrepancies in compute-optimal scaling of language models. In: The Thirty-eighth Annual Conference on Neural Information Processing Systems (2024)
22. Vorontsov, E., et al.: A foundation model for clinical-grade computational pathology and rare cancers detection. Nat. Med. **30**(10), 2924–2935 (2024). https://doi.org/10.1038/s41591-024-03141-0
23. Wang, S., et al.: Annotation-efficient deep learning for automatic medical image segmentation. Nat. Commun. **12**(1), 5915 (2021). https://doi.org/10.1038/s41467-021-26216-9
24. Yang, H., et al.: Multi-modal vision-language model for generalizable annotation-free pathology localization and clinical diagnosis. arXiv preprint arXiv:2401.02044 (2024)
25. Yang, J., et al.: MEDMNIST V2 - a large-scale lightweight benchmark for 2D and 3D biomedical image classification. Sci. Data **10**, 41 (2023). https://doi.org/10.1038/s41597-022-01721-8
26. Zhai, X., Kolesnikov, A., Houlsby, N., Beyer, L.: Scaling vision transformers. In: 2022 IEEE/CVF Conference on Computer Vision and Pattern Recognition (CVPR), pp. 1204–1213 (2022)

27. Zhang, K., et al.: A generalist vision-language foundation model for diverse biomedical tasks. Nat. Med. **30**, 3129–3141 (2024). https://doi.org/10.1038/s41591-024-03185-2
28. Zhang, S., Metaxas, D.: On the challenges and perspectives of foundation models for medical image analysis. Med. Image Anal. **91**, 102996 (2024). https://doi.org/10.1016/j.media.2023.102996
29. Zhang, S., et al.: BiomedCLIP: a multimodal biomedical foundation model pretrained from fifteen million scientific image-text pairs. arXiv preprint arXiv:2303.00915 (2023)
30. Zhou, H.Y., Chen, X., Zhang, Y., Luo, R., Wang, L., Yu, Y.: Generalized radiograph representation learning via cross-supervision between images and free-text radiology reports. Nat. Mach. Intell. **4**(1), 32–40 (2022). https://doi.org/10.1038/s42256-021-00425-9
31. Zhou, H.Y., Lu, C., Chen, C., Yang, S., Yu, Y.: A unified visual information preservation framework for self-supervised pre-training in medical image analysis. IEEE Trans. Pattern Anal. Mach. Intell. **45**(7), 8020–8035 (2023). https://doi.org/10.1109/TPAMI.2023.3234002
32. Zhou, H.Y., et al.: A transformer-based representation-learning model with unified processing of multimodal input for clinical diagnostics. Nat. Biomed. Eng. **7**(6), 743–755 (2023). https://doi.org/10.1038/s41551-023-01045-x
33. Zhou, Y., et al.: A foundation model for generalizable disease detection from retinal images. Nature **622**(7981), 156–163 (2023). https://doi.org/10.1038/s41586-023-06555-x

From Pathology to Radiology: Evaluating the Applicability of Pathology Foundation Models

Hyun Yang[1,2], Sumin Jung[2], and Jin Tae Kwak[1,2(✉)]

[1] School of Electrical and Electronic Engineering, Korea University,
Seoul, Republic of Korea
{buddygus,jkwak}@korea.ac.kr
[2] DeepClue Inc., Seoul, Republic of Korea
sumin00716@deepclue.io

Abstract. Foundation models, large-scale models pre-trained on vast and diverse datasets, effectively capture general knowledge and demonstrate strong robustness to a wide range of downstream tasks. However, the applicability of such foundation models to different domains and modalities has not been well studied. In this work, we introduce PathRadX, a cross-modal framework to evaluate the applicability of pathology foundation models to various classification tasks in radiology images. The framework integrate modality adaptation and task-specific classification strategies for efficient and effective cross-modal adaptation. Our findings highlight the cross-modal potential of pathology pre-trained models, demonstrating their applicability to radiology imaging tasks and paving the way for developing more efficient and accurate diagnostic tools.

Keywords: Foundation models · Radiology images · Deep learning.

1 Introduction

With the recent advancements in deep learning and the increasing accessibility of large-scale models, foundation models, trained on massive datasets, have gained considerable attention [1]. These models often leverage multi-modal datasets and are primarily trained through self-supervised or supervised learning methods, capturing generalizable knowledge that enables adaptation to a variety of tasks with minimal task-specific adjustments or even in a zero-shot setting. This diminishes the need for training models from scratch for each task, facilitating their application to specific tasks with limited data and annotations. As foundation models demonstrate strong adaptability across diverse tasks and datasets, domain-specific foundation models have been widely explored. Notable examples include GPT-3 [2] and BERT [3] in natural language processing, CLIP [4] in vision-language learning, and the Segment Anything Model (SAM) [5] in vision segmentation. The emergence of these models has led to significant breakthroughs across multiple domains, including language understanding, computer vision, and multi-modal learning.

In medical imaging, several foundation models have been developed using various imaging modalities, such as RadFM [6], a radiology foundation model aimed to integrate text input with 2D or 3D medical images, SPAD-Nets [7], trained on 55 public medical image datasets which contain both 2D and 3D images, and RETFound [8], a foundation model for abnormality detection from retinal images. Recently, there have been substantial advances in pathology foundation models, such as UNI [9], Virchow [10], and RudolphV [11]. These pathology foundation models are particularly notable due to their ability to process high-resolution images that preserve fine-grained and complex structural features. Previous studies [12] have shown that foundation models are applicable across different modalities. However, the adaptability and applicability of pathology foundation models to other medical imaging modalities remain largely unexplored.

Herein, we hypothesize that pathology foundation models possess sufficient generalization ability to be effectively adapted to other medical imaging modalities, such as radiology images. Given their ability to capture fine-grained structural and morphological features from pathology images, these models may be able to extract meaningful representations from radiology images with minimal modifications. To test this hypothesis, we propose PathRadX, a cross-modal framework that adapts a pathology foundation model to tasks specific to radiology. PathRadX involves modality adaptation and task-specific classification strategies, enabling an efficient and effective adaptation of a pathology foundation model to various radiology-specific tasks. We systematically evaluate the performance of PathRadX on three radiology datasets and compare it with three foundation models from other imaging modalities to assess its effectiveness in cross-modal medical image analysis.

Our main contributions can be summarized as below:

- We propose to adapt pathology foundation models to resolve radiology imaging tasks, leveraging their strength in analyzing fine-grained structures.
- We investigate an effective adaptation framework, referred to as PathRadX, incorporating domain adaptation and task-specific classification strategies, for pathology foundation models that requires minimal modifications and fine-tuning.
- We demonstrate that pathology foundation models exhibit strong generalization capabilities to radiology images, achieving promising results in cross-modal adaptation.

2 Methodology

PathRadX is a cross-modal framework designed to evaluate the applicability of a pathology foundation model across various tasks in radiology imaging. Given

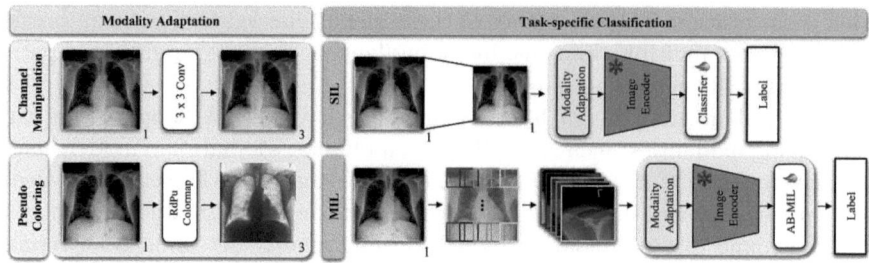

Fig. 1. Overview of the proposed PathRadX framework. To make gray-scale radiology images compatible with a pathology foundation model, we convert them into three-channel images using two different methods: pseudo-coloring with the RdPu colormap and channel manipulation via a single convolution layer. To resolve specific tasks in radiology, we employ two classification strategies: single instance learning and multiple instance learning.

a set of radiology datasets $D = \{D_i\}_{i=1}^N$ corresponding to tasks $T = \{T_i\}_{i=1}^N$, we employ a pre-trained pathology foundation model $\mathcal{P}$ to conduct each task T_i. To fairly assess the generalization ability of $\mathcal{P}$, we freeze all its pre-trained weights. However, there are modality differences, in which pathology images are typically three-channel color images and radiology images are generally one-channel gray-scale images, and class label differences across tasks. To address these challenges, we introduce a trainable adaptation layer $\mathcal{A}$ to bridge the modality gap and a task-specific classification layer $\mathcal{C}$ to accommodate different classification objectives. Consequently, we construct a set of pathology-based models for radiology imaging tasks, defined as $\mathcal{M} = \{\mathcal{M}_i\}_{i=1}^N$, where $\mathcal{M}_i = \mathcal{A}_i \circ \mathcal{P} \circ \mathcal{C}_i$ is tailored for each task T_i.

2.1 Pathology Foundation Model

Among various pathology foundation models, we adopt UNI to perform radiology imaging tasks. UNI [9] is a state-of-the-art (SOTA) framework designed for pathology image analysis, which is pre-trained on Mass-100K, a large-scale dataset comprising over 100 million image patches from 100,000 whole slide images (WSIs). It is built based upon Vision Transformer Large (ViT-L) [13] architecture and utilizes DINOv2 [14] for self-supervised learning. UNI has demonstrated SOTA performance on 34 pathology image tasks, highlighting its effectiveness in capturing fine-grained structural features and complex spatial relationships in pathology images.

2.2 Modality Adaptation

Pathology foundation models are tailored to process and analyze pathology images, typically containing three-channel RGB images. UNI is not an exception. However, these models are not directly applicable to gray-scaled radiology images. To mitigate this modality discrepancy, we propose two adaptation strategies to effectively utilize pathology foundation models for radiology imaging tasks: 1) Channel manipulation; 2) Pseudo coloring.

Channel Manipulation. The straightforward approach to convert a single-channel gray-scale image into a three-channel format involves the use of a convolution layer. We apply a 3 × 3 convolution layer with a stride of 1 and zero-padding of 1, preserving the spatial dimension of inputs while increasing the number of channels to match the expected number of channels of UNI.

Pseudo Coloring. We also colorize gray-scale radiology images by mapping gray-scale intensity values into a three-channel color space. Pathology images are typically stained with hematoxylin and eosin (H&E), where cell nuclei appear purplish-blue and cytoplasm appear pink. To mimic the color patterns of H&E stained pathology images, we apply a pseudo-colorization technique using the red-purple colormap from OpenCV [15], which exhibits a color gradient from light pink/red to deep purple (Fig. 2).

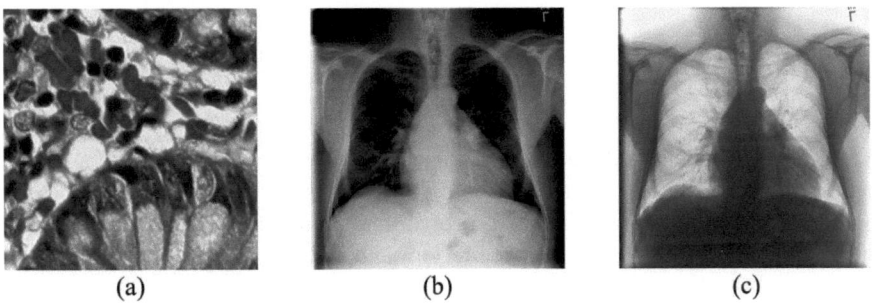

Fig. 2. Example of application of pseudo-coloring method on chest X-ray. (a) Example of a histopathology image, (b) Original chest X-ray image from the RSNA pneumonia dataset, and (c) Pseudo-colored chest X-ray image

2.3 Task-specific Classification

To adapt a pathology foundation model $\mathcal{P}$, i.e., UNI, for radiology imaging tasks, we explore two classification strategies: 1) single instance learning; 2) multiple instance learning. Given a radiology image, $\mathcal{P}$ employs one of these two classification strategies to extract a feature representation and predict the final classification label.

Single Instance Learning. In this approach, we size radiology images to 224× 224 pixels to match the input spatial size of UNI. The resized images are then fed into the encoder of UNI to extract a feature representation. Subsequently, the feature representation is passed through trainable linear classification layers to produce the final classification output.

Multiple Instance Learning. In pathology image analysis, whole-slide images (WSIs) are extremely large, typically around 50,000 × 50,000 pixels, as they are acquired through high-resolution scanning. Due to their size, a direct processing of WSIs is computationally intractable. Alternatively, WSIs are divided into a number of disjoint image patches, each of which is independently processed to extract feature representations. These feature representations are then aggregated by employing multiple instance learning (MIL) approaches, generating a single representative feature representation that is used for predicting the final classification output. Recent studies have shown that pathology foundation models, equipped with MIL, achieve the SOTA performance in WSI analysis. To extend MIL to radiology images, we adopt a similar approach by cropping radiology images into multiple overlapping patches of size 224 × 224 pixels with an overlap of 80 pixels or 24 pixels, considering the original image size, to generate corresponding feature representations, in which each patch image is independently processed by UNI. The extracted feature representations are then aggregated using attention-based MIL (AB-MIL) [16], which utilizes adaptive attention weights to aggregate feature representations, to predict the final label.

3 Experiments

3.1 Datasets

We employ three different radiology datasets to evaluate adaptability of a pathology foundation model to radiology images. These datasets consist of X-ray images of different organs. Table 1 provides a summary of these datasets and the training, validation, and test splits for each dataset.

MURA. [17] is a large dataset of musculoskeletal radiographs containing 40,561 X-ray images from 14,863 studies, containing 9,045 normal and 5,818 abnormal radiographic studies of the upper extremity including the shoulder, humerus, elbow, forearm, wrist, hand, and finger. As 556 images were used as the hidden test set during the challenge, 40,005 X-ray images were used in this study, with diverse spatial sizes up to 512 × 512 pixels.

RSNA Pneumonia Dataset ($RSNA_P$). [18] was originally made in public during RSNA Pneumonia Detection Challenge held in 2018, containing 26,686 images with the spatial size of 1024 × 1024 pixels each. We use this dataset for a binary classification task to determine the presence of pneumonia.

MedFMC ($MedFMC_C$). [19] contains 22,349 images in total, covering 5 different modalities, X-ray, pathology, endoscopy, digital camera, and retinography. Among them, we use 4,848 Chest X-ray images to conduct thoracic abnormality classification. Each image has a spatial size of 1024 × 1024 pixels.

Table 1. Overview of radiology datasets with detailed characteristics. Unlike other datasets, MURA has variable size of images.

Dataset	Organ	Image size	Train	Validation	Test	Total
MURA	musculoskeletal X-ray	$\leq 512 \times 512$	36,808	1,500	1,697	40,005
$RSNA_P$	chest X-ray	$1,024 \times 1,024$	21,348	2,668	2,668	26,684
$MedFMC_C$	chest X-ray	$1,024 \times 1,024$	3,878	485	485	4,848

3.2 Comparative Experiments

To evaluate the effectiveness of the proposed PathRadX framework, we conduct comparative experiments using three different radiology foundation models: 1) SAM-Med2D; 2) MedCLIP; 3) BiomedCLIP. Following the standard practice in radiology foundation models, each model processes an input radiology image through its image encoder to extract a feature representation, which is fed into trainable linear layers for classification.

SAM-Med2D. SAM has set a benchmark for segmenting instances in natural images through its prompt-based approach. As a variant of SAM, SAM-Med2D [20] was built by fine-tuning SAM with approximately 4.6 million medical images and 19.7 million corresponding masks. It has been shown that SAM-Med2D outperforms existing segmentation methods on nine MICCAI2023 datasets.

MedCLIP. MedCLIP [22] is a vision-language model designed for medical image analysis, pre-trained on a large dataset of chest X-ray images paired with clinical text, such as CheXpert [23] and MIMIC-CXR [24]. By learning to associate visual features with medical terminology, it has demonstrated strong performance, particularly in chest X-ray image analysis.

BiomedCLIP. BiomedCLIP extends MedCLIP by focusing on broader biomedical applications. It utilizes PubMedBERT [25] as a text encoder and a ViT model as an image encoder. Pre-trained on a large biomedical dataset called PMC-15M [26], including histopathology images, radiology images, and clinical notes, BiomedCLIP enhances the interaction between biomedical images and text, improving performance across various medical imaging tasks.

3.3 Implementation and Training Details

The image encoders of each foundation model were frozen, and only the classifier weights were updated during training. Each model was trained for 100 epochs with OneCycleLR [27] scheduler (initial learning rate 0.001) and the Adam optimizer, with 0 weight decay and beta values (β_1, β_2) set to 0.9 and 0.999. All experiments were conducted on a single NVIDIA RTX A6000 GPU. To evaluate and train the models, we adopt binary cross-entropy loss.

4 Results and Discussions

Table 2 demonstrates the performance of each foundation model on radiology imaging datasets based on three metrics: accuracy (Acc), F1 score, and AUC score. Leveraging UNI with AB-MIL and channel-manipulation, we achieved the highest scores across all evaluation metrics for MURA and $RSNA_P$, outperforming others by ≤6.4% Acc, ≤0.081 F1, and ≤0.073 AUC for MURA and ≤4.0% Acc, ≤0.243 F1, and ≤0.054 AUC for $RSNA_P$. Meanwhile, on $MedFMC_C$, Med-CLIP obtained the highest Acc of 76.29% and AUC of 0.784 and SAM-Med2D obtained the best F1 of 0.852. The proposed method was inferior to these models by 3.3% Acc, 0.040 F1, and 0.016 AUC.

To provide an in-depth understanding of the classification results, we computed and examined confusion matrices for each model (Figure 3). In a close-up investigation of these confusion matrices, we found that two of the three datasets ($RSNA_P$ and $MedFMC_C$) were highly imbalanced. The imbalance particularly affects the performance on $MedFMC_C$. SAM-Med2D predicted all samples as positive. MedCLIP had much lower specificity than the proposed method, leading to a high rate of false positives. Overall, the proposed method generally demonstrated more balanced and reliable classification performance.

Moreover, we evaluated the proposed method with differing strategies for modality adaptation and task-specific classification (Table 3). Except $MedFMC_C$, the combination of AB-MIL and channel-manipulation provided high performance across three metrics. On $MedFMC_C$, equipped with SIL, the proposed method achieved much higher scores for both modality adaptation strategies, even outperforming MedCLIP. These results suggest that optimizing modality adaptation and task-specific classification strategies could further improve the performance.

Table 2. Results of three radiology imaging tasks. CM: channel manipulation, UNI: a pathology foundation model; MIL: multiple instance learning.

Models	MURA			$RSNA_P$			$MedFMC_C$		
	Acc(%)	F1	AUC	Acc(%)	F1	AUC	Acc(%)	F1	AUC
PathRadX (CM ∘ UNI ∘ MIL)	**77.31**	**0.777**	**0.861**	**83.70**	**0.576**	**0.863**	72.99	0.812	0.768
SAM-Med2D	70.94	0.726	0.788	79.72	0.333	0.809	74.23	**0.852**	0.757
MedCLIP	74.96	0.696	0.844	81.90	0.488	0.834	**76.29**	0.848	**0.784**
BiomedCLIP	75.25	0.700	0.835	81.52	0.417	0.820	75.26	0.851	0.762

Furthermore, future studies could systematically evaluate the potential of pathology foundation models on radiology tasks through broader comparisons. As an extension of our current findings, evaluating the proposed framework against models pretrained on large-scale natural image datasets may help further

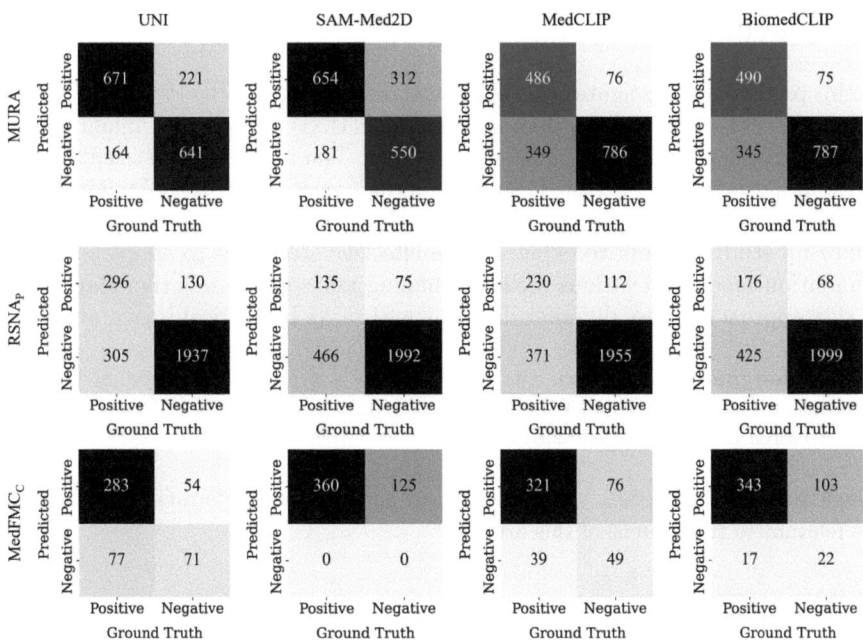

Fig. 3. Visualization of confusion matrices of evaluation results.

Table 3. Efffect of different modality adaptation and task-specific classification strategies on the performance of a pathology foundation model. SIL: single instance learning, MIL: multiple instance learning, CM: channel manipulation, PC: and pseudo-coloring.

Strategy				MURA			$RSNA_P$			$MedFMC_C$		
SIL	MIL	CM	PC	Acc(%)	F1	AUC	Acc(%)	F1	AUC	Acc(%)	F1	AUC
✓		✓		**77.84**	0.752	0.848	81.33	0.357	0.836	79.18	0.870	**0.819**
✓			✓	77.61	0.744	0.849	80.88	0.348	0.844	**79.59**	**0.873**	0.807
	✓	✓		77.31	**0.777**	**0.861**	**83.70**	**0.576**	**0.863**	72.99	0.812	0.768
	✓		✓	77.67	0.753	0.839	81.30	0.540	0.829	77.73	0.852	0.781

validate the robustness of PathRadX for radiology imaging tasks. Additionally, incorporating models such as CTransPath [28] or Lunit [29] into follow-up experiments would enable exploration of cross-domain strategies, potentially improving the effectiveness of pathology-specific models when applied to radiology tasks.

5 Conclusions

In this paper, we investigate four different strategies within the PathRadX framework to leverage a pathology foundation model (UNI) on radiology imaging tasks in comparison to radiology foundation models. The results suggest that pathology foundation models are promising to conduct various radiology imaging tasks. However, the specific strategies for adaptation needs further investigation. In a follow-up study, we aim to explore more effective strategies to adapt pathology foundation models to various medical imaging tasks and assess the adaptability of this approach across different domains and tasks in medical image analysis.

Acknowledgment. This work was supported by a grant of the National Research Foundation of Korea (NRF) (No. RS-2025-00558322) and Ministry of Health and Welfare of Korea (No. RS-2023-00266130).

Disclosure of Interests. The authors have no competing interests to declare that are relevant to the content of this article.

References

1. Bommasani, R., et al.: On the opportunities and risks of foundation models. arXiv preprint arXiv:2108.07258 (2021)
2. Brown, T., et al.: Language models are few-shot learners. Adv. Neural. Inf. Process. Syst. **33**, 1877–901 (2020)
3. Devlin, J., Chang, M.W., Lee, K., Toutanova, K.: BERT: pre-training of deep bidirectional transformers for language understanding. In: Proceedings of the 2019 Conference of the North American Chapter of the Association for Computational Linguistics: Human Language Technologies, Volume 1 (long and short papers), pp. 4171–4186 (2019)
4. Radford, A., et al.: Learning transferable visual models from natural language supervision. In: International Conference on Machine Learning, pp. 8748–8763 (2021)
5. Kirillov, A., et al.: Segment anything. In: Proceedings of the IEEE/CVF International Conference on Computer Vision 2023, pp. 4015–4026 (2023)
6. Wu, C., Zhang, X., Zhang, Y., Wang, Y., Xie, W.: Towards generalist foundation model for radiology by leveraging web-scale 2D&3D medical data. arXiv preprint arXiv:2308.02463 (2023)
7. Luo, L., et al.: Building universal foundation models for medical image analysis with spatially adaptive networks. arXiv preprint arXiv:2312.07630 (2023)
8. Zhou, Y., Chia, M.A., Wagner, S.K., et al.: A foundation model for generalizable disease detection from retinal images. Nature **622**, 156–163 (2023). https://doi.org/10.1038/s41586-023-06555-x
9. Chen, R.J., et al.: a general-purpose self-supervised model for computational pathology. arXiv preprint arXiv:2308.15474 (2023)
10. Vorontsov, E., et al.: Virchow: a million-slide digital pathology foundation model. arXiv preprint arXiv:2309.07778 (2023)
11. Dippel, J., et al.: RudolfV: a foundation model by pathologists for pathologists. arXiv preprint arXiv:2401.04079 (2024)

12. Huix, J.P., Ganeshan, A.R., Haslum, J.F., Söderberg, M., Matsoukas, C., Smith, K.: Are natural domain foundation models useful for medical image classification?. In: Proceedings of the IEEE/CVF Winter Conference on Applications of Computer Vision, pp. 7634–7643 (2024)
13. Dosovitskiy, A., et al.: An image is worth 16x16 words: transformers for image recognition at scale. arXiv preprint arXiv:2010.11929 (2020)
14. Oquab, M., et al.: Dinov2: learning robust visual features without supervision. arXiv preprint arXiv:2304.07193 (2023)
15. Bradski, G., Kaehler, A.: OpenCV: open source computer vision library. Proc. Intel Technol. J. **2000**, 1–11. https://opencv.org
16. Ilse, M., Tomczak, J., Welling, M.: Attention-based deep multiple instance learning. In: International Conference on Machine Learning, pp. 2127–2136 (2018)
17. Rajpurkar, P., et al.: Mura: large dataset for abnormality detection in musculoskeletal radiographs. arXiv preprint arXiv:1712.06957 (2017)
18. RSNA pneumonia detection challenge. https://kaggle.com/competitions/rsna-pneumonia-detection-challenge
19. Wang, D., et al.: A real-world dataset and benchmark for foundation model adaptation in medical image classification. Sci. Data **10**(1), 574 (2023)
20. Cheng, J., et al.: SAM-MED2D. arXiv preprint arXiv:2308.16184 (2023)
21. Wang, X., Peng, Y., Lu, L., Lu, Z., Bagheri, M., Summers, R.M.: CHESTX-RAY8: hospital-scale chest x-ray database and benchmarks on weakly-supervised classification and localization of common thorax diseases. In: Proceedings of the IEEE Conference on Computer Vision and Pattern Recognition, pp. 2097–2106 (2017)
22. Wang, Z., Wu, Z., Agarwal, D., Sun, J.: MedClip: contrastive learning from unpaired medical images and text. In: Proceedings of the Conference on Empirical Methods in Natural Language Processing, vol. 2022, p. 3876 (2022)
23. Irvin, J., et al.: CheXpert: a large chest radiograph dataset with uncertainty labels and expert comparison. In: Proceedings of the AAAI Conference on Artificial Intelligence, vol. 33, no. 01, pp. 590–597 (2019)
24. Johnson, A.E., et al.: MIMIC-CXR-JPG, a large publicly available database of labeled chest radiographs. arXiv preprint arXiv:1901.07042 (2019)
25. Gu, Y., et al.: Domain-specific language model pretraining for biomedical natural language processing. ACM Trans. Comput. Healthc. (HEALTH) **3**(1), 1–23 (2021)
26. Zhang, S., et al.: BioMedClip: a multimodal biomedical foundation model pretrained from fifteen million scientific image-text pairs. arXiv preprint arXiv:2303.00915 (2023)
27. Smith, L.N., Topin, N.: Super-convergence: very fast training of neural networks using large learning rates. In: Artificial Intelligence and Machine Learning for Multi-Domain Operations Applications, vol. 11006, pp. 369–386 (2019)
28. Xiyue, W., et al.: Transformer-based unsupervised contrastive learning for histopathological image classification. Med. Image Anal. **81**, 102559 (2022)
29. Mingu, K., Heon, S., Seonwook, P., Donggeun, Y., Sérgio, P.: Benchmarking self-supervised learning on diverse pathology datasets. In: Proceedings of the IEEE/CVF Conference on Computer Vision and Pattern Recognition, pp. 3344–3354 (2023)

Pathology Foundation Models Are Scanner Sensitive: Benchmark and Mitigation with Contrastive *ScanGen* Loss

Gianluca Carloni(✉)[iD], Biagio Brattoli[iD], Seongho Keum[iD], Jongchan Park[iD], Taebum Lee[iD], Chang Ho Ahn[iD], and Sergio Pereira[iD]

Lunit, Seoul, Republic of Korea
gianluca.carloni@lunit.io

Abstract. Computational pathology (CPath) has shown great potential in mining actionable insights from Whole Slide Images (WSIs). Deep Learning (DL) has been at the center of modern CPath, and while it delivers unprecedented performance, it is also known that DL may be affected by irrelevant details, such as those introduced during scanning by different commercially available scanners. This may lead to *scanner bias*, where the model outputs for the same tissue acquired by different scanners may vary. In turn, it hinders the trust of clinicians in CPath-based tools and their deployment in real-world clinical practices. Recent pathology Foundation Models (FMs) promise to provide better domain generalization capabilities. In this paper, we benchmark FMs using a multi-scanner dataset and show that FMs still suffer from scanner bias. Following this observation, we propose *ScanGen*, a contrastive loss function applied during task-specific fine-tuning that mitigates scanner bias, thereby enhancing the models' robustness to scanner variations. Our approach is applied to the Multiple Instance Learning task of Epidermal Growth Factor Receptor (EGFR) mutation prediction from H&E-stained WSIs in lung cancer. We observe that *ScanGen* notably enhances the ability to generalize across scanners, while retaining or improving the performance of EGFR mutation prediction.

Keywords: Computational Pathology · Scanner Generalization · Foundation Models

1 Introduction

Digital pathology is catalyzing a new era in the field of histopathology by transforming how clinicians interact with and analyze tissue samples. Furthermore, it enables computational pathology (CPath) over whole slide images (WSIs), a key component for deriving novel and quantitative insights from WSIs. This brings

---

G. Carloni and B. Brattoli—Shared first authorship.

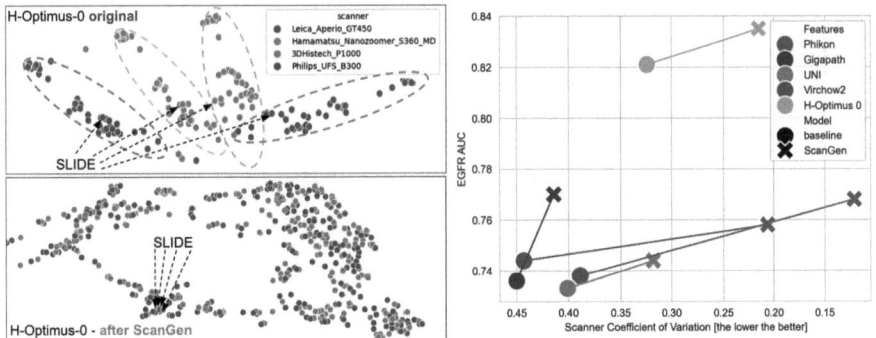

Fig. 1. Left: Existing pathology Foundation Models suffer from scanner bias. We show a 2D projection of WSIs (dots) from 4 scanners (colors) using UMAP based on H-Optimus-0 embeddings before and after *ScanGen*. "SLIDE→" indicates the slides of the same specimen from different scanners. Right: *ScanGen* improves both generalization and performance.

new capabilities to clinical practice, such as prediction of genetic alterations [11,22], tumor micro-environment analysis [16], or survival prediction [18].

Digital pathology requires the digitization of tissue slides into WSIs using slide scanners. While these scanners must capture the underlying tissue accurately, different devices may result in differences in image characteristics [5]. For example, differences in optical systems may impact image sharpness, variations in light sources can affect contrast and introduce artifacts, and proprietary software (e.g. for data compression) can affect image processing. To implement CPath-based tools in clinical settings, it is crucial to address the constraints specific to different scanners. Since different scanners are used worldwide, with some being predominant in certain regions (e.g., Hamamatsu in Asia [17]), any performance issues associated with a particular scanner could lead to regional disparities in patient care.

Recently, the introduction of foundation models (FMs) provides a base upon which CPath tools are developed. However, while FMs promise better generalization capabilities, they may still encode several biases, such as the slide identity [24] and hospital environments [9]. The present paper delves into another crucial aspect of CPath, the digital scanners used during image acquisition, offering empirical evidence that FMs are sensitive to the type of digital scanner employed for reading tissue specimens. Since CPath FMs are typically used as frozen extractors with no access to the input images, alleviating this bias with traditional stain normalization methods (e.g., Reinhard [19], Macenko [14]) or DL ones (e.g., StainGAN [21], StainNet [10]) is not trivial. Scanner bias results in weakened generalization capabilities for FM-based CPath models, posing challenges in their broader application. In particular, scanner bias means that the model output will change based on which scanner has been used to read the specimen despite the content being the same, which can significantly hinder trust in CPath-based tools and, consequently, slower clinical adoption.

To address this shortcoming, we propose *ScanGen*, a contrastive loss function that mitigates scanner bias, thereby enhancing the model's robustness across different scanners. Our evaluation involves testing various state-of-the-art FMs using a benchmark set that comprises images of the same specimen captured from six commercially available scanners at 40× magnification and one scanner at 20× magnification. Additionally, our approach is applied to the task of detecting Epidermal Growth Factor Receptor (EGFR) mutations in lung cancer, a prevalent and clinically significant endeavor in the CPath space [2,25].

In summary, this paper introduces three contributions: 1) a contrastive loss function aimed at reducing scanner bias in Multiple Instance Learning (MIL), 2) an assessment of FMs in relation to scanner generalization, and 3) application of the proposed method to EGFR mutation prediction in lung cancer, a clinically relevant task. The results demonstrate promising improvements in predictive performance and generalization (Fig. 1), underscoring the potential of our methodologies in clinical settings.

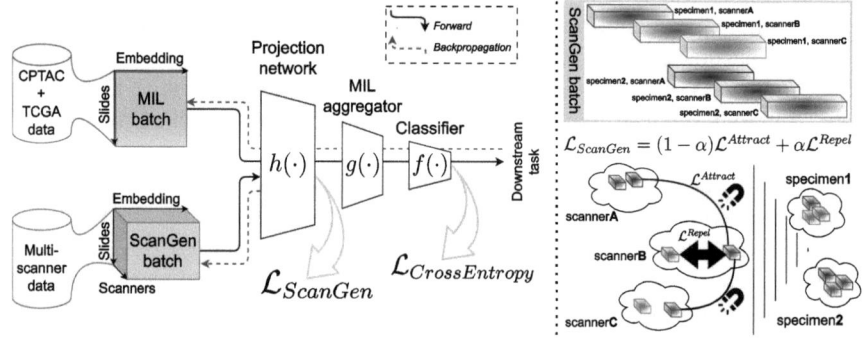

Fig. 2. Our method. Left: Data are projected to a new embedding space where scanner bias is removed via our contrastive *ScanGen* loss. *ScanGen* is trained on paired WSIs of the same specimen acquired with different scanners and can be easily integrated with any existing method by acting before the MIL aggregator. Right: *ScanGen* attracts the representations of same-specimen different-scanner pairs, while repelling same-scanner different-specimen ones. This favors specimen-based clusters to scanner-based ones, thus fostering generalization across scanners in downstream applications.

2 Method

2.1 Scanner Generalization Metric for MIL

Due to the extremely large size of WSIs, MIL is usually employed for classification or regression tasks. In MIL, WSIs are partitioned into P patches and the

$FM(\cdot)$ is used as a pretrained, frozen feature extractor to obtain an embedding vector $x_i = FM(p_i)$ per each patch p_i. Finally, an aggregator function $g(\cdot)$ combines all patches and a classifier $f(\cdot)$ produces the final prediction (e.g., EGFR mutation) as $\hat{y} = f(g(x_1, \ldots, x_P))$. In this context, the scanner bias relates to how different two predictions $\hat{y}_m^{s_1}$ and $\hat{y}_m^{s_2}$ are from the same specimen m but different scanners s_1 and s_2. For this reason, we measure the intensity of scanner bias as the average coefficient of variation (CoV) between predictions of the same specimen m across S multiple scanners, defined as

$$\text{CoV} = \frac{1}{M}\sum_{m=1}^{M}\left(\frac{\sigma(\hat{y}_m^{s_1}, \hat{y}_m^{s_2}, \ldots, \hat{y}_m^{s_S})}{|\mu(\hat{y}_m^{s_1}, \hat{y}_m^{s_2}, \ldots, \hat{y}_m^{s_S})|}\right), \quad (1)$$

with M number of specimens, where σ and μ are, respectively, the standard deviation and average across the logit predictions of the same specimen acquired with different scanner hardware. CoV is more appropriate than simple standard deviation because it allows for meaningful comparisons across sources with different scales. Lower CoV means better agreement in predictions, indicating the model is less scanner-sensitive.

2.2 Proposed Framework and *ScanGen* Loss

As Fig. 2 illustrates, to reduce the scanner bias in the embedding space, we first map the patch embeddings x to a new scanner-unbiased embedding space using a projection network $h(\cdot)$, which is trained as follows. Given two specimens m and n and two scanners s_1 and s_2, we rework the contrastive loss function [7] and define our *ScanGen* loss as

$$\mathcal{L}_{ScanGen} = (1-\alpha)\underbrace{(d(h(x_m^{s_1}), h(x_m^{s_2})))^2}_{\mathcal{L}^{Attract}} + \alpha\underbrace{(\max(0, r - d(h(x_m^{s_1}), h(x_n^{s_1}))))^2}_{\mathcal{L}^{Repel}} \quad (2)$$

where $d(a,b) = 1 - \frac{a \cdot b}{|a| \cdot |b|}$ is the cosine distance function, $\mathcal{L}^{Attract}$ is the attraction term encouraging samples from the same specimen but different scanners to be close in the embedding space, $\mathcal{L}^{Repel}$ is the repulsion term pushing samples from the same scanner but different specimens apart, r is the radius, and α is a hyper-parameter governing the weighted combination of those terms. The MIL aggregator is then applied: $g(\cdot) = g(h(x_1), h(x_2), \ldots, h(x_P))$.

The final loss is composed of the downstream task loss, in our case EGFR prediction using standard cross-entropy (CE), and our *ScanGen* loss:

$$\mathcal{L} = \frac{1}{|A|}\sum_{a \in A}\mathcal{L}_{CE}(\hat{y}_a, y_a) + \lambda \cdot \frac{1}{|B|}\sum_{\substack{m,n \in B \\ m \neq n}}\sum_{\substack{u,v \in S \\ u \neq v}}\mathcal{L}_{ScanGen}(x_m^{s_u}, x_m^{s_v}, x_n^{s_u}; h) \quad (3)$$

where A and B are independent sets of labeled samples for the downstream task and the *ScanGen* task, respectively, S is the number of scanners, and λ is a hyper-parameter to control the balance between those two terms.

2.3 Implementation

Within a shared model, two different data flows occur at each forward pass during training (see Fig. 2-left). On the one hand, a regular MIL batch sampled from an EGFR-labeled data pool enters the projection-aggregation-classification cascade, thus producing the logits for standard CE loss computation.

On the other hand, a batch of embeddings identified by specimen and scanner names is fed into the projection network $h(\cdot)$. There, the *ScanGen* loss first calculates a distance matrix using pairwise cosine similarity between embeddings and constructs matrices to identify, within the batch, pairs of samples from the same scanner or with the same specimen name. These matrices are used to compute the attraction and repulsion losses, which are weighted and combined on the basis of the α parameter. The final loss is averaged over valid pairs, ensuring that only relevant pairs (i.e., either same-specimen different-scanners or same-scanner different-specimen) contribute to the loss calculation. For fair comparison, also the baseline models where *ScanGen* is not enabled utilize the projection network $h(\cdot)$, which is learned, together with $g(\cdot)$ and $f(\cdot)$, under the CE loss alone.

We developed our code in PyTorch. We implement $h(\cdot)$ as a multi-layer perceptron (MLP) with three layers and hidden dimension between 48 and 96 depending on the FM features. Regarding $g(\cdot)$, we use the global average pooling (GAP) as a simple MIL model, where patch-level features are averaged to produce one embedding per WSI: $X = \frac{1}{P}\sum_{i=1}^{P} x_i$, followed by an MLP classifier. Moreover, Table 2 shows an adaptation to some modern MIL aggregators, for which we followed their official implementation. In our experiments, we found that α of 0.12 to 0.21, r of 0.90 to 1.10, and λ of 0.5 to 1.0 are appropriate.

3 Results

3.1 Datasets

Two publicly available datasets, CPTAC [4] and TCGA [1]), for a total of 1,144 WSIs, were used to train the EGFR branch. Note that only non-small cell lung cancer specimens were selected. Our multi-scanner dataset, on the other hand, is composed of 323 specimens, each tested for EGFR mutation using PCR and NGS assay as well as scanned using six devices: Leica Aperio AT2, Leica Aperio GT450, Hamamatsu Nanozoomer S360MD, 3DHistech P1000, Philips UFS B300, Roche Ventana DP200. The specimens were scanned at 40× magnification. The specimens were also scanned with Leica Aperio GT450 at 20× magnification. In total, the multi-scanner dataset is composed of $323 * 7 = 2,261$ WSIs. Figure 4-right shows different scans of the same specimen using the six different scanners. We split this dataset into three splits while preserving class balance and specimen-level split (i.e., WSIs of the same specimen were grouped). As a result, 54 specimens were used to train the *ScanGen* branch, 45 specimens were used as validation set for tuning both EGFR branch and *ScanGen* branch, and 224 specimens were used as held-out test set for external evaluation of downstream task performance (EGFR AUC) and scanner agreement (CoV).

3.2 Evaluating Scanner Generalization

We assess FMs on our multi-scanner dataset, presenting both qualitative and quantitative outcomes. To analyze the bias for each FM visually, for each WSI in the test set, we average all patch embeddings to produce a single WSI embedding, and project the full test set in 2D using UMAP [15] with cosine metric and default parameters. Figure 3 illustrates that each FM is affected by scanner bias, as evidenced by the clustering and separation of samples according to the scanner rather than the same specimen. The only exception is UNI [3], which, according to the 2D projection, seems to be unaffected by this bias; however, our quantitative results paint a different picture, and even UNI largely benefits from our *ScanGen* loss.

In Table 1, we benchmark five state-of-the-art FMs according to our new metrics for evaluating generalization, CoV over scanners and CoV over magnifications, and show *ScanGen*'s mitigation effect. Our expectation is that *ScanGen* would reduce the CoV, while avoiding hindrance on the EGFR downstream task. Our findings indicate that incorporating *ScanGen* projection significantly enhances the ability to generalize across scanners (lower CoV). More importantly, the predictive performance for EGFR is not hindered; rather, it actually improves (higher AUC) for each embedding type. Additionally, according to our findings, the *ScanGen* loss also helps mitigate a disagreement in predictions when different magnifications are used to read the slide.

A similar result is shown in Fig. 1-right, where each model is compared on CoV and EGFR AUC, with and without *ScanGen* loss. While CoV significantly improves for each FM, we also see a boost in EGFR AUC.

Table 1. Evaluating scanner generalization for available foundation models on held-out testset. Coefficient of EGFR Variation (CoV) is shown across six scanners and two magnifications (40×, 20×).

Embedding	Use *ScanGen*	EGFR AUC [↑]	Scanner CoV [↓]	Magnification CoV [↓]
Phikon [6]	No	0.738	0.389	0.158
	Yes	**0.768**	**0.121** (-68.9%)	**0.053** (-66.4%)
UNI [3]	No	0.733	0.401	0.116
	Yes	**0.744**	**0.318** (-20.7%)	**0.089** (-23.3%)
Virchow [26]	No	0.744	0.443	0.125
	Yes	**0.758**	**0.206** (-53.5%)	**0.077** (-38.4%)
Gigapath [23]	No	0.736	0.450	**0.174**
	Yes	**0.770**	**0.414** (-8.00%)	0.193 (+10.9%)
H-Optimus-0 [20]	No	0.821	0.324	0.117
	Yes	**0.835**	**0.215** (-33.6%)	**0.108** (-7.70%)

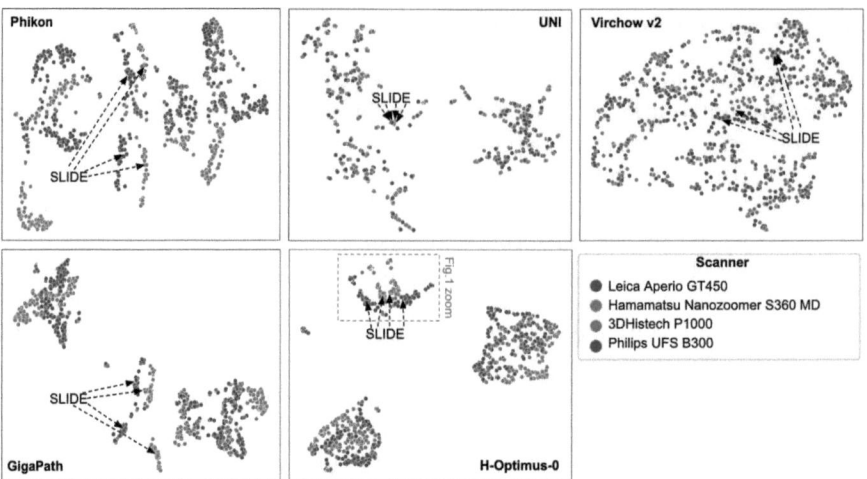

Fig. 3. Our multi-scanner dataset projected to 2D using various FMs. Each dot is a WSI embedding from the test set by averaging FM embeddings across patches and projecting them using UMAP. The color indicates the scanner. We selected 4 out of 7 scanners to reduce clutter. A single slide is indicated with "SLIDE→" to emphasize the location of the slides of the same specimen from different scanners in each plot.

3.3 *ScanGen* with Limited Scanners

This section presents an ablation study evaluating the impact of varying the number of scanner types used for *ScanGen* loss during model training. Figure 4 highlights the gain of CoV compared to the baseline. Each model is tested on the same external test set containing all six available scanner sources. It examined scenarios where different combinations of two to six scanner types were utilized for the training process. The plot indicates the monotonic correlation between the number of scanner types and the scanner generalization performance: when more scanner types are used for the *ScanGen* loss, the generalization across scanners increases. Performance improvement converges at five scanners. Nevertheless, even when only three scanner types are used for training, the scanner generalization on six test scanners significantly improves by 27%.

This ablation study also allowed us to investigate which scanner combinations led to the highest or lowest CoV improvement over baseline. We counted how often a certain scanner was present in the best (i.e., lowest CoV) and worst combinations. Doing this for all ablation scenarios from two to five scanners, we found that Leica Aperio GT450 and the Hamamatsu scanners were *always* in the best performing configurations, while the Leica Aperio AT2 and GT450 pair was the one most often associated with the worst performing configurations. This suggests that using a diverse set of scanners for *ScanGen* loss leads to the best generalization, while using similar scanners, like the two Leica Aperio, does not benefit scanner robustness much.

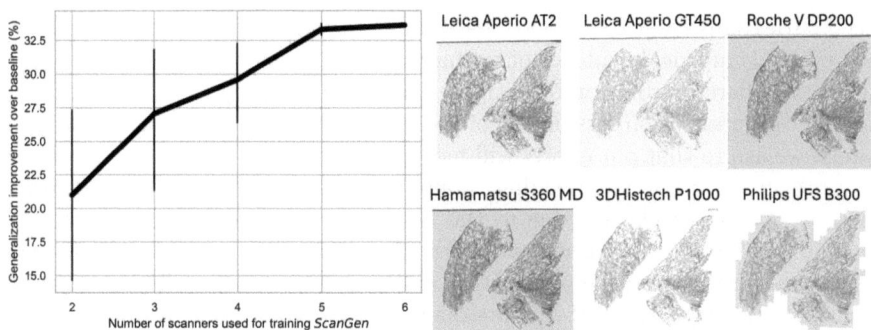

Fig. 4. Left: Scanner agreement improves with the number of scanners used for training *ScanGen*. Vertical bars indicate the standard deviation across scanner combinations. Right: Examples of the same specimen digitized using different scanners.

3.4 Extension to Other MIL Methods

We also showcase our method's performance on recent MIL techniques like AB-MIL [8], DS-MIL [13], and SlotMIL [12]. Table 2 summarizes our findings with H-Optimus-0 features. Our *ScanGen* loss consistently reduced CoV, boosting agreement by 11% to 33%, and maintained or surpassed the baseline AUC.

Table 2. Integrating our *ScanGen* loss into other MIL approaches. We report the AUC and scanner CoV values obtained on the held-out test set.

MIL approach	GAP		AB-MIL [8]		DS-MIL [13]		SlotMIL [12]	
	AUC	CoV [↓]	AUC	CoV [↓]	AUC	CoV [↓]	AUC	CoV [↓]
EGFR	0.821	0.324	0.828	0.251	**0.834**	0.265	0.822	0.338
EGFR + *ScanGen*	**0.835**	**0.215**	**0.835**	**0.224**	0.828	**0.214**	**0.826**	**0.227**

4 Discussion

This study demonstrates that current FMs for CPath are sensitive to scanners, impacting the model results. In the global market, scanner distribution is uneven. Leica and Philips lead in America, 3DHistech challenges Leica in Europe, and Hamamatsu is dominant in Asia [17]. Differences in device performance may affect quality or care between Asia, the U.S., and Europe. To build real-world CPath tools on a global scale, we need to think about generalization across hospitals in different geographical regions, which requires more robust predictive models.

Our research is essential in emphasizing a current problem stemming from the swift adoption of FMs in numerous CPath applications. We first gathered a benchmark comprising six widely used scanners available in the market. Then, we presented empirical evidence of scanner bias across various cutting-edge FMs. Finally, we offered a solution to mitigate the bias in existing FMs by integrating a scanner generalization loss on top of the standard MIL downstream task. Ultimately, we aspire that our efforts will motivate future FM developers to address this problem during model training. Nonetheless, training an FM is beyond the reach of most research labs due to the necessity of tens of thousands of samples and significant computational effort. Consequently, our initial focus is on demonstrating the presence of the issue and suggesting a cost-effective solution to enhance the robustness of existing open-source FMs against scanner bias.

Disclosure of Interests. The authors have no competing interests to declare that are relevant to the content of this article.

References

1. Albertina, B., et al.: The cancer genome atlas lung adenocarcinoma collection (TCGA-LUAD) (2016)
2. Campanella, G., et al.: H&E-based computational biomarker enables universal EGFR screening for lung adenocarcinoma. arXiv preprint arXiv:2206.10573 (2022)
3. Chen, R.J., et al.: Towards a general-purpose foundation model for computational pathology. Nat. Med. **30**(3), 850–862 (2024)
4. (CPTAC) National Cancer Institute: The clinical proteomic tumor analysis consortium lung adenocarcinoma collection (CPTAC-LUAD) (version 12) [data set] (2018). https://doi.org/10.7937/K9/TCIA.2018.PAT12TBS
5. Duenweg, S.R., et al.: Whole slide imaging (WSI) scanner differences influence optical and computed properties of digitized prostate cancer histology. J. Pathol. Inform. **14** (2023)
6. Filiot, A., et al.: Scaling self-supervised learning for histopathology with masked image modeling. medRxiv (2023)
7. Hadsell, R., Chopra, S., LeCun, Y.: Dimensionality reduction by learning an invariant mapping. In: 2006 IEEE Computer Society Conference on Computer Vision and Pattern Recognition (CVPR'06), vol. 2, pp. 1735–1742. IEEE (2006)
8. Ilse, M., Tomczak, J., Welling, M.: Attention-based deep multiple instance learning. In: International Conference on Machine Learning, pp. 2127–2136 (2018)
9. de Jong, E.D., Marcus, E., Teuwen, J.: Current pathology foundation models are unrobust to medical center differences. arXiv preprint arXiv:2501.18055 (2025)
10. Kang, H., et al.: StainNet: a fast and robust stain normalization network. Front. Med. **8**, 746307 (2021)
11. Kather, J.N., et al.: Pan-cancer image-based detection of clinically actionable genetic alterations. Nature cancer **1**(8), 789–799 (2020)
12. Keum, S., Kim, S., Lee, S., Lee, J.: Slot-mixup with subsampling: a simple regularization for WSI classification. arXiv preprint arXiv:2311.17466 (2023)
13. Li, B., Li, Y., Eliceiri, K.W.: Dual-stream multiple instance learning network for whole slide image classification with self-supervised contrastive learning. In: Proceedings of the IEEE/CVF Conference on Computer Vision and Pattern Recognition, pp. 14318–14328 (2021)

14. Macenko, M., et al.: A method for normalizing histology slides for quantitative analysis. In: 2009 IEEE International Symposium on Biomedical Imaging: from Nano to Macro, pp. 1107–1110. IEEE (2009)
15. McInnes, L., Healy, J., Melville, J.: UMAP: uniform manifold approximation and projection for dimension reduction. arXiv preprint arXiv:1802.03426 (2018)
16. Park, S., et al.: Artificial intelligence-powered spatial analysis of tumor-infiltrating lymphocytes as complementary biomarker for immune checkpoint inhibition in non-small-cell lung cancer. J. Clin. Oncol. **40**(17), 1916–1928 (2022)
17. Pinto, D.G., Bychkov, A., Tsuyama, N., Fukuoka, J., Eloy, C.: Exploring the adoption of digital pathology in clinical settings-insights from a cross-continent study. medRxiv (2023)
18. Ramanathan, V., Pati, P., McNeil, M., Martel, A.L.: Ensemble of prior-guided expert graph models for survival prediction in digital pathology. In: International Conference on Medical Image Computing and Computer-Assisted Intervention, pp. 262–272 (2024)
19. Reinhard, E., Adhikhmin, M., Gooch, B., Shirley, P.: Color transfer between images. IEEE Comput. Graphics Appl. **21**(5), 34–41 (2001)
20. Saillard, C., et al.: H-optimus-0 (2024). https://github.com/bioptimus/releases/tree/main/models/h-optimus/v0
21. Shaban, M.T., Baur, C., Navab, N., Albarqouni, S.: StainGAN: stain style transfer for digital histological images. In: 2019 IEEE 16th International Symposium on Biomedical Imaging (ISBI 2019), pp. 953–956. IEEE (2019)
22. Teichmann, M., Aichert, A., Bohnenberger, H., Ströbel, P., Heimann, T.: End-to-end learning for image-based detection of molecular alterations in digital pathology. In: International Conference on Medical Image Computing and Computer-Assisted Intervention, pp. 88–98 (2022)
23. Xu, H., et al.: A whole-slide foundation model for digital pathology from real-world data. Nature, 1–8 (2024)
24. Yun, J., Hu, Y., Kim, J., Jang, J., Lee, S.: EXAONEPath 1.0 patch-level foundation model for pathology. arXiv preprint arXiv:2408.00380 (2024)
25. Zhao, D., et al.: High accuracy epidermal growth factor receptor mutation prediction via histopathological deep learning. BMC Pulmonary Med. **23**(1) (2023)
26. Zimmermann, E., et al.: Virchow2: scaling self-supervised mixed magnification models in pathology. arXiv preprint arXiv:2408.00738 (2024)

Improved Training Sample Efficiency and Inter-device Generalizability in Optical Coherence Tomography Fluid Segmentation via Foundation Models

Yusuke Kikuchi[1]([✉]) [iD], Matthew McLeod[1], Eric Gros[1], Maxime Usdin[1], Ali Boushehri[2] [iD], Julia Cluceru[1] [iD], Yaniv Cohen[2], Yvonna Li[2], and Qi Yang[1]

[1] Genentech, Inc., South San Francisco, CA, USA
{kikuchi.yusuke,mcleod.matthew.mm1,gros.eric,usdin.maxime, cluceru.julia,yang.qi}@gene.com
[2] F. Hoffmann-La Roche AG, Basel, Switzerland
{ali.boushehri,yaniv.cohen,yvonna.li}@roche.com

Abstract. Accurate segmentation of fluid regions in optical coherence tomography (OCT) images is crucial for ophthalmologic diagnosis and treatment monitoring. However, automated segmentation models face two key challenges: high annotation costs and limited generalization across OCT devices. We investigate foundation models to address these challenges, evaluating a domain-specific OCT foundation model trained using the SimCLR method and an adapted Segment Anything Model 2 (SAM2). Through experiments with datasets ranging from 50 to 2,500 images and cross-device validation, we show that foundation models outperform ImageNet-pretrained models in small data regimes. In cross-device evaluation, both foundation models demonstrated superior generalization. Our findings indicate that foundation models significantly reduce annotation requirements and enhance cross-device adaptability, lowering development costs and accelerating deployment of OCT fluid segmentation solutions.

Keywords: Optical Coherence Tomography · Foundation Models · Medical Image Segmentation · Sample Efficiency · Generalizability

1 Introduction

In the treatment and management of retinal exudative diseases such as neovascular age-related macular degeneration (nAMD) and diabetic macular edema (DME), the accurate measurement of fluid in the retina is crucial for making treatment decisions and predicting the treatment outcome [4,12]. Optical Coherence Tomography (OCT) is a non-invasive imaging modality, operating at micrometer-scale resolution, which enables the detailed visualization of retinal structure and pathology, particularly the accumulation of fluid in different retinal layers.

The process of manually annotating fluid regions in OCT scans is labor-intensive and time-consuming. Expert graders, typically experienced ophthalmologists or trained technicians, must meticulously delineate fluid boundaries across numerous B-scan slices. This labor-intensive process is prohibitive in real world clinical settings [2]. The challenge necessitates the development of automated systems that can segment the fluid regions in OCT scans. Previous works have shown promising results using deep learning [14]. However, collecting the manual annotations is still a significant bottleneck in the development pipeline.

Additionally, the field faces a challenge in achieving reliable inter-device generalizability, as OCT imaging characteristics can vary significantly across different manufacturer platforms [5]. For instance, Spectralis (Heidelberg Engineering, Inc.) and Cirrus (Carl Zeiss Meditec, Inc.) systems produce images with distinct resolution profiles and noise characteristics (Fig. 1). Mitigating the degradation and improving model generalizability is highly desirable because device-specific training requires a costly process of collecting the manual annotations for each device.

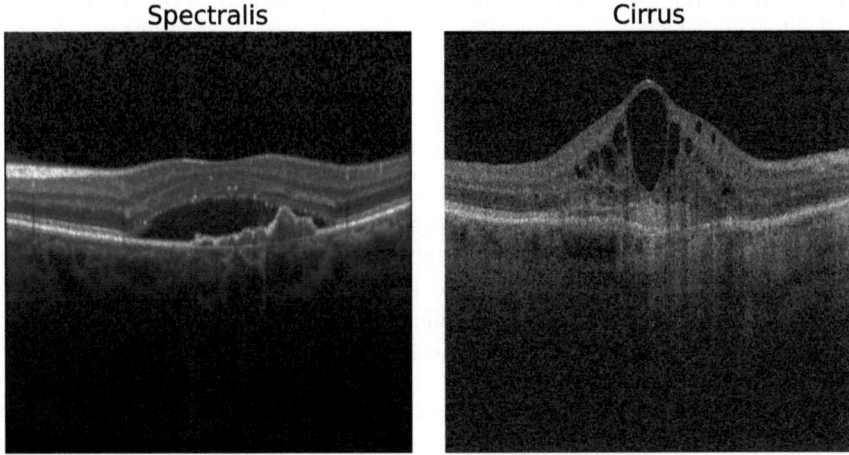

Fig. 1. Sample images from the Spectralis (Heidelberg Engineering) and the Cirrus (Carl Zeiss Meditec) OCT devices. Both images were resized to 512 by 512 pixels from the original size.

A foundation model is a model pretrained on a large heterogeneous dataset typically using self-supervised learning methods to acquire the general-purpose representation. The large scale pretraining of foundation models often eases the requirements on the size of the task-specific training set and exhibits better generalizability to unseen data [1]. In this study, we evaluate two distinct approaches of foundation model-based OCT fluid segmentation on the challenges in the training size requirement (sample efficiency) and inter-device generalization. The first approach utilizes an internally developed foundation model trained using the SimCLR method [3], specifically trained on a large corpus of OCT B-scans to capture domain-specific features. The second approach leverages Segment Anything Model 2 [11] architecture, a publicly available foundation model originally designed for general-purpose image segmentation.

While most foundation model research focuses primarily on absolute performance metrics, sample efficiency and out-of-distribution generalization are two critical aspects that have been less frequently and thoroughly addressed. Our analysis is centered around these two aspects in the task of OCT fluid segmentation. Specifically, we address the following questions through carefully designed experiments.

Q1. (Sample Efficiency) How much labeled data for training segmentation models can be saved by leveraging foundation models compared to the standard approach without foundation models?

Q2. (Inter-device Generalizability) What is the extent of out-of-distribution generalizability of foundation models compared to the generalizability of the standard approach without foundation models? Additionally, how does the performance of foundation models compare to the performance of the in-domain standard approach?

2 Methods

2.1 OCT Fluid Segmentation

Our investigation centers on the challenging task of pixel-level multi-class fluid segmentation in OCT images, specifically focusing on three distinct fluid types: Intraretinal fluid (IRF), subretinal fluid (SRF), and pigment epithelial detachment (PED). To conduct the sample efficiency and inter-device generalizability experiments, we used two different datasets.

- D_S: Spectralis dataset. The dataset consists of OCT B-scans acquired using Spectralis (Heidelberg Engineering) devices in AVENUE clinical trial in nAMD (NCT02484690) and BOULEVARD clinical trial in DME (NCT02699450). The dataset was split into training, validation, and holdout testing. There are 2,568 B-scans for training, 458 B-scans for validation, and 371 B-scans for holdout testing.
- D_C: Cirrus dataset. The dataset consists of OCT B-scans acquired using Zeiss (Carl Zeiss Meditec) devices in HARBOR clinical trial in nAMD (NCT06411288). HARBOR is a clinical trial in nAMD. The dataset was also split into training, validation, and holdout testing. There are 735 B-scans for training, 126 B-scans for validation, and 99 B-scans for holdout testing.

All the splits were done at the patient level to avoid any potential data leakage. The annotation was done by trained graders and the quality was reviewed by a senior grader. More detailed protocol is available in [10]. The roles of the datasets are described in 2.4.

2.2 In-Domain Foundation Model Development

SimCLR Foundation Model. Following the SimCLR framework [3], our model was trained on a large dataset of unlabeled OCT B-scans, which include clinical studies in nAMD, DME, and geographic atrophy (GA). The OCT scans were taken by devices manufactured by Heidelberg Engineering, Inc., Carl Zeiss Meditec, Inc., and Topcon Corporation, Inc. The training data were curated to exclude any patients whose data are in the validation and the holdout sets of D_S and D_C.

The training set consists of 765,586 OCT volumes from 9,509 patients (27% nAMD, 37% DME, and 36% GA). The device distribution is 70% Spectralis, 26% Cirrus, and 4% Topcon. To prepare the training data for SimCLR pretraining, individual B-scans were extracted from OCT volumes. Because the B-scans taken at off-central locations are less likely to contain pathological biomarkers and the appearance is similar, only the B-scans from the middle 50% on the transversal direction were included. As a result, the training set consists of 18,916,869 B-scans.

A ResNet50 [6] model was used as the backbone model. The SimCLR training followed the original framework including the input image size of 224 by 224 pixels. The batch size was 6,000 and the number of epochs was 40. The weights of ResNet50 were initialized with ImageNet weights.

Adapting Foundation Model for Segmentation. The SimCLR model only consists of the encoder part and therefore we need to convert it to a segmentation model by adding a decoder part. Based on the popularity and the proven performance of the U-Net architecture [13] in medical image segmentation, we chose to convert the ResNet50 model to a ResUNet [7] architecture. In fine-tuning on the segmentation task, the decoder part was randomly initialized and the entire model was trained end to end.

Regarding the training of ResUNet, we made two observations: 1) Some target masks, typically IRF, could be very small in the size of 224 by 224 pixels. Concretely, the majority of the IRF masks were less than 0.2% area of the image, which is approximately 100 pixels in the image size of 224 by 224 pixels. 2) At the beginning of the training, the weights of the encoder are already optimized after the SimCLR pretraining while the decoder part is freshly initialized. To address 1), we used a larger image size of 512 by 512 pixels for fine-tuning. For 2), a different learning rate for the encoder and the decoder was included in the hyper parameter search (the options of learning rate are 1e-4 and 1e-5 for the encoder and 1e-3 and 1e-4 for the decoder). Due to the limitation in time and computational resources, we used the default parameters for the following parameters: equally weight Dice and focal loss for the loss function, AdamW [8] for the optimizer, linear warm up and cosine annealing [9] for the learning rate scheduler, and the batch size of 20.

2.3 SAM2-Based Approach

Segment Anything Model 2 (SAM2) is a foundation model for general-purpose image segmentation. It was trained on a large dataset of annotated natural videos [11]. We adapted SAM2 for the OCT fluid segmentation task by closely following the SAM2-UNet architecture and implementation [15]. The SAM2-UNet architecture is comprised of the SAM2 encoder and a convolutional decoder where the intermediate feature representation from different resolutions are connected to the decoder by skip connections, which makes the whole architecture a U-Net [13] shape. To facilitate efficient training, the authors of SAM2-UNet employed a parameter-efficient fine-tuning approach and structure loss that combines intersection-over-union loss and cross-entropy loss that highlights the boundaries [15]. The two observations about the image size and the learning rate for the encoder and decoder were also examined for SAM2-UNet as well. For the image size, the default image size for SAM2 of 1024 by 1024 pixels was used because our concern was about small image size. For the learning rate, the trainable parameters in SAM2-UNet are the adaptors and the decoder. Because all of them are randomly initialized, we used the same learning rate for the encoder and the decoder. We tuned the learning rate (1e-3 and 1e-4) and the kernel size in the structure loss (31, 41, and 51). Again, due to the computational resource limitation, we fixed the following parameters: AdamW [8] for the optimizer, cosine annealing [9] for the learning rate scheduler, and the batch size of 4.

The model structure of adapting the foundation models and the ImageNet pretrained model is shown in Fig. 2.

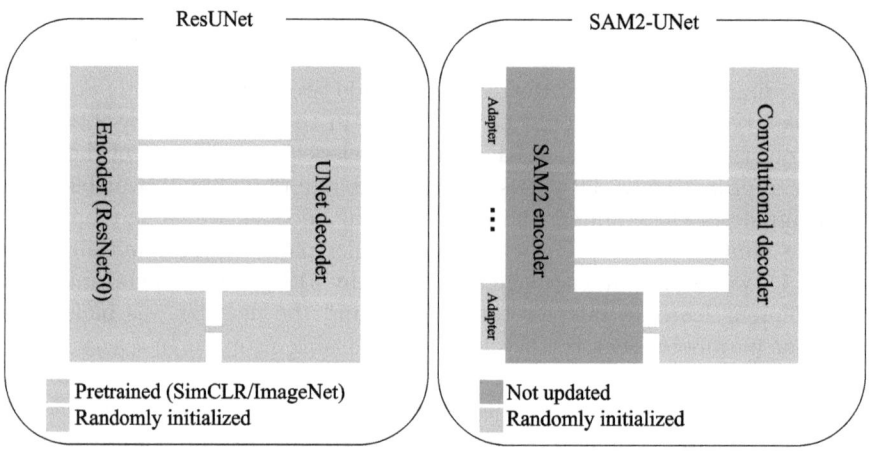

Fig. 2. Model structure of adapting the foundation models and the ImageNet pretrained model. ResUNet (left) and SAM2-UNet (right).

2.4 Experimental Design

Sample Efficiency Evaluation. The sample efficiency experiments were designed to evaluate model performance across different training data regimes. The following five distinct dataset proportions of the training set of D_S were used: 100%, 40% (= 1,028 ≈ 1,000 B-scans), 20% (=514 ≈ 500 B-scans), 4% (=103 ≈ 100 B-scans), and 2% (= 52 ≈ 50 B-scans). The validation and the holdout sets were kept the same regardless of the training set size. To ensure the training set contains representative cases, we stratified the training set by the total fluid volume and then randomly sampled the cases from each stratum for five distinct data portions. In hyper parameter tuning, we ran the training process five times with different random seeds to average out the randomness. The models with the hyper parameter combination with the lowest validation loss averaged over the random seeds was selected to be evaluated on the holdout test set.

Inter-device Generalizability Assessment. The evaluation of inter-device generalizability focused on zero-shot performance of models trained on images taken on Spectralis devices when applied to images taken with Cirrus devices. The models trained on 100% of D_S training set that were selected for the holdout testing were evaluated on the holdout testing set of D_C.

2.5 Evaluation Metrics and Baseline Models

The final model performance on the holdout set was evaluated using classwise Dice score excluding the background class. Note that the Dice score when both the target and the prediction are empty is defined to be 1.

To highlight the added values from foundation models, we compare the performance of the foundation models with the baseline models. We chose the ResUNet model with the encoder initialized with ResNet50 ImageNet weights as the baseline model. The training details are the same as the ResUNet models with SimCLR ResNet50 encoder. Hereafter, we refer to the ResUNet model with the initial weights pretrained on our internal data using the SimCLR as SimCLR ResUNet and the ResUNet model with the initial weights from ImageNet as ImageNet ResUNet. Also, we collectively call SimCLR ResUNet and SAM2-UNet as foundation model-based approaches.

We also added ImageNet ResUNet trained on D_C to the comparison of the inter-device generalizability experiments. This addition enables us to compare the performance of foundation model-based approaches with a standard approach in D_C when training data is available.

2.6 Implementation Details

All the experiments were conducted on NVIDIA A100 GPUs with 40GB of memory. The SimCLR ResNet50 was trained using 16 GPUs. The ResUNet models and SAM2-UNet models were trained using a single GPU.

3 Results

3.1 Sample Efficiency Analysis

Our evaluation of sample efficiency revealed several significant findings across different training data regimes. The results demonstrated a clear advantage of foundation model-based approaches over traditional training methods, particularly in low-data scenarios. We look at the results from the two comparisons: 1) ImageNet ResUNet vs. foundation model-based approaches and 2) SAM2-UNet vs. SimCLR ResUNet. For the first comparison, the foundation model-based approaches consistently outperformed the ImageNet ResUNet and showed more robust performance as the training data size decreased as shown in Fig. 3. Examining the same figure, the foundation model-based approaches already reached the performance level at 100% with 40% training data and there are slight drops at 20% from 40%. On the other hand, the ImageNet ResUNet performance starts to drop at 40% compared to the performance at 100%. This would mean that the training size needed to fine-tune the foundation models is in between 500 and 1,000 samples and the ImageNet ResUNet needs more than 1,000 samples to fine-tune. For the second comparison within foundation model-based approaches, Fig. 3 shows that the SAM2-UNet outperformed the SimCLR ResUNet with 4% to 100% of the training data in all fluid types while the relative rank changes with 2% of training data. The reason why SAM2-UNet exhibits a more acute performance drop than SimCLR ResUNet from 4% to 2% is not entirely clear without further experiments. However, a potential factor is the differences in the pretraining dataset. Because SAM2 was trained on natural images, it may need more training data to adapt to the OCT images than SimCLR ResUNet, which was trained on OCT images.

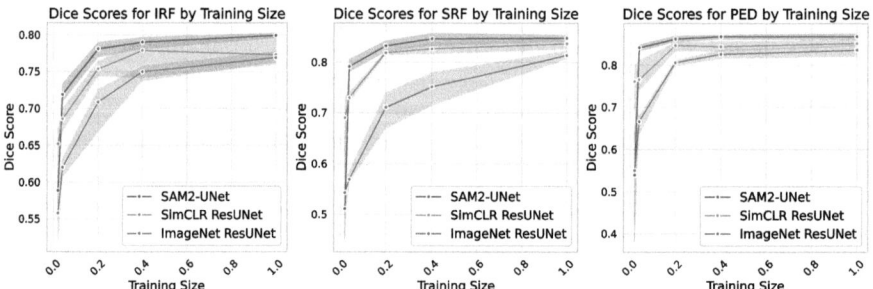

Fig. 3. Sample efficiency results. The mean holdout test Dice score for 5 random seeds is shown. The shaded area represents the highest and lowest mean Dice score out of the 5 random seeds.

3.2 Effect of Image Size and Learning Rate

Using a different image size in fine-tuning from pretraining is not standard practice. However, our results suggest that the fine-tuning process could help the models adapt to the larger image size.

Additionally, we report that a smaller learning rate was always preferred for the encoder part than the decoder part regardless of the initial encoder weights in the hyper parameter tuning of ResUNet.

3.3 Inter-device Generalizability

The models evaluated on the holdout testing set of D_S in the sample efficiency analysis were evaluated on the holdout testing set of D_C in a zero-shot way. The zero-shot performance was compared to the supervised ImageNet ResUNet model trained on D_C. The holdout test results are shown in Table 1. Overall, the foundation model-based approaches showed superior performance compared to the zero-shot and in-domain ImageNet ResUNet models. Among the foundation model-based approaches, SimCLR ResUNet and SAM2-UNet are similar in SRF and PED while SAM2-UNet showed strength in IRF. Looking at the predicted masks shown in Fig. 4, the foundation model-based approaches respect the topology of the fluid regions, which implies those models learned it in the source domain in a generalizable way. The results indicate the strong generalization capability of the foundation models across different OCT devices.

Table 1. Inter-device generalizability results. The mean holdout test Dice score over 5 random seeds on the holdout testing set of D_C. The model name follows the form "{model architecture} on {fine-tuning dataset}". The values in parentheses are the minimum and maximum Dice score out of the 5 random seeds.

Model	Dice score (IRF)	Dice score (SRF)	Dice score (PED)
SimCLR ResUNet on D_S	0.71 (0.68, 0.75)	0.71 (0.67, 0.73)	0.64 (0.60, 0.70)
SAM2-UNet on D_S	0.76 (0.73, 0.79)	0.72 (0.70, 0.74)	0.64 (0.60, 0.69)
ImageNet ResUNet on D_S	0.69 (0.66, 0.71)	0.45 (0.39, 0.57)	0.42 (0.34, 0.49)
ImageNet ResUNet on D_C	0.70 (0.65, 0.72)	0.55 (0.48, 0.67)	0.59 (0.57, 0.61)

4 Discussion and Conclusion

4.1 Key Findings and Implications

Our investigation into foundation model-based approaches for OCT fluid segmentation has revealed several significant findings. The superior performance of foundation model-based approaches suggests a potential for significant reduction

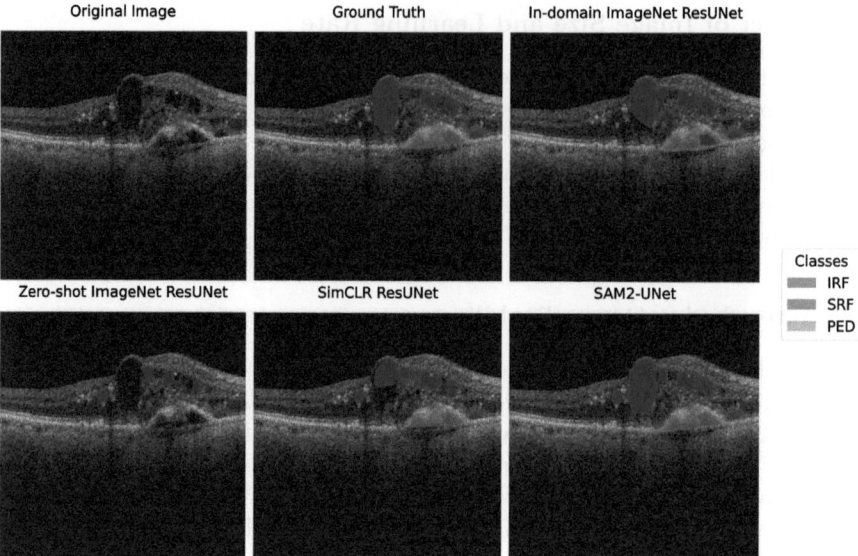

Fig. 4. A representative case highlighting the qualitative differences between different approaches. The in-domain ImageNet ResUNet predicted SRF inside PED, which violates the topology of the fluid regions. The Zero-shot ImageNet ResUNet missed a large portion of IRF and completely missed PED. The overall quality was improved in foundation model-based approaches although SimCLR ResUNet missed part of the IRF region and SAM2-UNet oversegmented.

in annotation requirements for future applications, which enables less time and cost for collecting data and ultimately leads to shorter development time.

In this study, we included two foundation models with different pretraining datasets and methods. SAM2 showed a pronounced domain adaptability in OCT fluid segmentation and outperformed SimCLR ResUNet in most of the cases except the very small data regime with about 50 B-scans for training. Even though the model size is a magnitude smaller, SimCLR ResUNet showed the advantage of in-domain pretraining by following SAM2-UNet closely in most of the experimental settings.

It is also notable that the SimCLR ResUNet consistently outperformed the ImageNet ResUNet in all experimental settings, which solely came from the difference in the initial weights. This supports that in-domain pretraining is advantageous in the context of medical image analysis, which aligns with the findings in the previous works such as [16].

4.2 Limitations and Future Works

In our study, we evaluated the two foundation models in the context of OCT fluid segmentation with a selected adaptation strategy. To validate and generalize our findings more broadly, it is warranted to evaluate more foundation models and adaptation strategies on a diverse range of tasks.

References

1. Bommasani, R., et al.: On the opportunities and risks of foundation models. CoRR **abs/2108.07258** (2021). https://arxiv.org/abs/2108.07258
2. Breger, A., et al.: Supervised learning and dimension reduction techniques for quantification of retinal fluid in optical coherence tomography images. Eye **31**(8), 1212–1220 (2017)
3. Chen, T., Kornblith, S., Norouzi, M., Hinton, G.: A simple framework for contrastive learning of visual representations. In: International Conference on Machine Learning, pp. 1597–1607 (2020)
4. Gerendas, B.S., Prager, S., Deak, G., Simader, C., Lammer, J., Waldstein, S.M., Guerin, T., Kundi, M., Schmidt-Erfurth, U.M.: Predictive imaging biomarkers relevant for functional and anatomical outcomes during ranibizumab therapy of diabetic macular oedema. Br. J. Ophthalmol. **102**(2), 195–203 (2018)
5. Gomariz, A., Kikuchi, Y., Li, Y.Y., Albrecht, T., Maunz, A., Ferrara, D., Lu, H., Goksel, O.: Joint semi-supervised and contrastive learning enables domain generalization and multi-domain segmentation. Med. Image Anal. **103**, 103575 (2025)
6. He, K., Zhang, X., Ren, S., Sun, J.: Deep residual learning for image recognition. In: Proceedings of the IEEE Conference on Computer Vision and Pattern Recognition, pp. 770–778 (2016)
7. Jha, D., Riegler, M.A., Johansen, D., Halvorsen, P., Johansen, H.D.: Resunet++: an advanced architecture for medical image segmentation. IEEE Int. Symp. Multimedia (ISM), pp. 225–255 (2019)
8. Loshchilov, I., Hutter, F.: Decoupled weight decay regularization. In: International Conference on Learning Representations (2017). https://api.semanticscholar.org/CorpusID:53592270
9. Loshchilov, I., Hutter, F.: SGDR: stochastic gradient descent with warm restarts. In: 5th International Conference on Learning Representations, ICLR 2017, Toulon, France, April 24-26, 2017, Conference Track Proceedings. OpenReview.net (2017). https://openreview.net/forum?id=Skq89SCxx
10. Maunz, A., et al.: Accuracy of a machine-learning algorithm for detecting and classifying choroidal neovascularization on spectral-domain optical coherence tomography. J. Personalized Med. **11**(6) (2021). https://doi.org/10.3390/jpm11060524
11. Ravi, N., et al.: Sam 2: segment anything in images and videos. arXiv preprint arXiv:2408.00714 (2024)
12. Reiter, G.S., Mares, V., Leingang, O., Fuchs, P., Bogunovic, H., Barthelmes, D., Schmidt-Erfurth, U.: Long-term effect of fluid volumes during the maintenance phase in neovascular age-related macular degeneration: results from fight retinal blindness! Can. J. Ophthalmol. **59**(5), 350–357 (2024)
13. Ronneberger, O., Fischer, P., Brox, T.: U-Net: convolutional networks for biomedical image segmentation. In: Navab, N., Hornegger, J., Wells, W.M., Frangi, A.F. (eds.) MICCAI 2015. LNCS, vol. 9351, pp. 234–241. Springer, Cham (2015). https://doi.org/10.1007/978-3-319-24574-4_28

14. Schmidt-Erfurth, U., Waldstein, S.M.: A paradigm shift in imaging biomarkers in neovascular age-related macular degeneration. Prog. Retin. Eye Res. **50**, 1–24 (2016)
15. Xiong, X., et al.: Sam2-unet: segment anything 2 makes strong encoder for natural and medical image segmentation (2024). https://arxiv.org/abs/2408.08870
16. Zhou, Y., et al.: A foundation model for generalizable disease detection from retinal images. Nature **622**(7981), 156–163 (2023)

Taming Stable Diffusion for Computed Tomography Blind Super-Resolution

Chunlei Li[1], Yilei Shi[1], Haoxi Hu[2], Jingliang Hu[1], Xiao Xiang Zhu[3], and Lichao Mou[1](✉)

[1] MedAI Technology (Wuxi) Co. Ltd., Wuxi, China
lichao.mou@medimagingai.com
[2] Northeastern University, Boston, USA
[3] Technical University of Munich, Munich, Germany

Abstract. High-resolution computed tomography (CT) imaging is essential for medical diagnosis but requires increased radiation exposure, creating a critical trade-off between image quality and patient safety. While deep learning methods have shown promise in CT super-resolution, they face challenges with complex degradations and limited medical training data. Meanwhile, large-scale pre-trained diffusion models, particularly Stable Diffusion, have demonstrated remarkable capabilities in synthesizing fine details across various vision tasks. Motivated by this, we propose a novel framework that adapts Stable Diffusion for CT blind super-resolution. We employ a practical degradation model to synthesize realistic low-quality images and leverage a pre-trained vision-language model to generate corresponding descriptions. Subsequently, we perform super-resolution using Stable Diffusion with a specialized controlling strategy, conditioned on both low-resolution inputs and the generated text descriptions. Extensive experiments show that our method outperforms existing approaches, demonstrating its potential for achieving high-quality CT imaging at reduced radiation doses. Code is available at https://github.com/MedAITech/SD4CTSR.

Keywords: CT super-resolution · Stable Diffusion · fine-tuning

1 Introduction

Computed tomography (CT) is an indispensable imaging modality in modern healthcare, serving a crucial role in disease diagnosis, treatment planning, and clinical research [1]. The diagnostic value of CT images heavily depends on their spatial resolution, driving the continuous pursuit of high-resolution CT imaging techniques [2]. However, achieving high-resolution CT scans often requires increased radiation exposure, which poses potential health risks to patients [3]. This fundamental trade-off between image quality and radiation dose has motivated extensive research in CT super-resolution methods, aiming to generate high-resolution CT images from low-dose acquisitions [4].

C. Li and Y. Shi—Equal Contribution.

Recent years have witnessed remarkable progress in image super-resolution, primarily driven by deep learning approaches. Early efforts [5,7,8] typically rely on simulated training data with simplified, uniform degradation models, such as bicubic downsampling. Although these methods show promise in controlled settings, they often underperform when applied to real-world low-resolution images characterized by complex, unknown degradations. To address these limitations, blind image super-resolution methods [10–12,31] have gained increasing attention, particularly in medical imaging applications where degradation patterns can be highly variable and unpredictable. Recent work has demonstrated success in radiographs and MRI through blind super-resolution approaches that generalize across degradation types [22,23].

Despite these remarkable advances, the performance of learning-based super-resolution methods is inherently tied to the availability of high-quality training data. Unlike natural image domains with abundant data, medical imaging faces data scarcity due to privacy concerns and regulatory requirements, creating a barrier to developing robust super-resolution models.

Recently, large-scale pre-trained text-to-image diffusion models, particularly Stable Diffusion [17], have demonstrated astonishing capabilities in generating highly detailed images. Trained on vast collections of image-text pairs, these models have not only revolutionized image generation but have also shown promise results when adapted to various vision tasks, including image segmentation and enhancement [24–27]. This success motivates our investigation into leveraging Stable Diffusion for CT super-resolution, utilizing its rich prior knowledge of image structure and detail synthesis.

However, applying Stable Diffusion to CT super-resolution presents challenges. A fundamental obstacle is the domain gap between the model's training data—comprising internet-scale natural image-text pairs—and CT images. Direct application of Stable Diffusion risks synthesizing features characteristic of natural images, which is problematic given CT images' distinct properties in grayscale distributions, structural patterns, and texture characteristics.

To address this issue, we propose a novel framework that adapts Stable Diffusion for CT blind super-resolution. First, we employ a practical degradation model [12] to synthesize realistic low-quality images. Second, we utilize LLaVA-Med [21], a specialized medical vision-language model, to generate anatomical descriptions from these degraded images, which serve as semantic prompts for the subsequent super-resolution process. Third, we adapt Stable Diffusion by freezing its pre-trained layers and introducing a side-controlling strategy to generate high-resolution CT images conditioned on low-resolution inputs and text prompts. Our approach effectively bridges the gap between Stable Diffusion's general-purpose generative capabilities and specialized medical imaging requirements. Our main contributions are:

- We present the first attempt to adapt Stable Diffusion for medical image super-resolution.

- We propose a two-stage framework combining LLaVA-Med with Stable Diffusion, enabling text-guided super-resolution for medical imaging, along with a side-controlling approach to address domain shift.
- State-of-the-art performance on CT super-resolution tasks, surpassing existing methods.

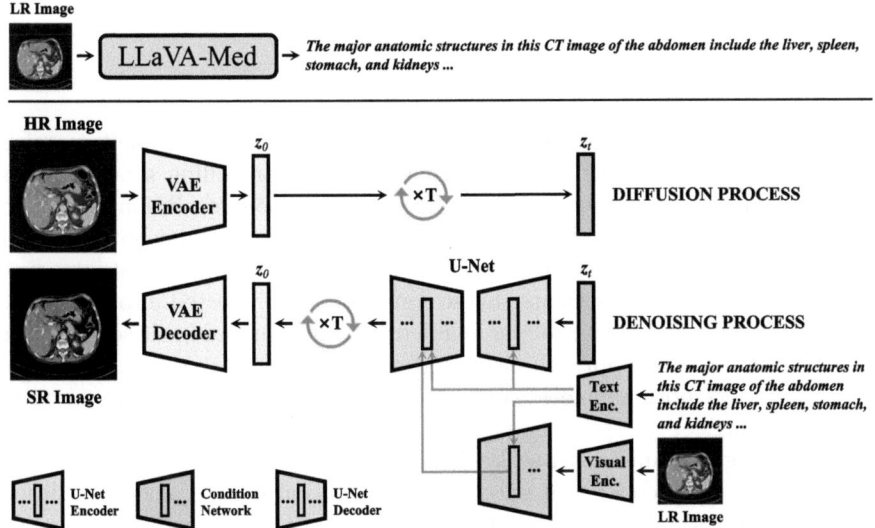

Fig. 1. Overview of our proposed framework. The pipeline consists of two stages: (1) text description generation from low-resolution images using LLaVA-Med, and (2) image super-resolution through the proposed side-controlling strategy, which integrates both the generated text prompts and low-resolution images into the Stable Diffusion model.

2 Method

As shown in Fig. 1, our framework consists of two main stages: text prompt generation and taming Stable Diffusion for super-resolution through the side-controlling strategy. The key innovation lies in fusing visual constraints from low-resolution inputs with semantic guidance derived from text prompts, enabling precise control over the pre-trained Stable Diffusion model.

2.1 Text Prompt Generation

To make full use of the generative potential of Stable Diffusion for super-resolution, we employ LLaVA-Med [21] to generate descriptive text prompts from low-dose CT images.

Given a low-resolution image x_{lr} and an instruction s (e.g., *Describe main anatomical structures present in the image*) as inputs, LLaVA-Med produces a corresponding textual description p, formulated as:

$$p = \mathcal{F}_{\text{LLaVA-Med}}(x_{lr}, s). \tag{1}$$

The resulting p provides valuable semantic information, particularly anatomical structures, which is crucial for medical imaging tasks. This description is then fed into Stable Diffusion as a text prompt, along with x_{lr}.

2.2 Side-Controlling

To bridge the domain gap between natural and medical images while leveraging the strength of Stable Diffusion, we propose a side-controlling strategy. This approach synergizes text prompts (see Sect. 2.1) and low-resolution images to provide complementary guidance during generation. The textual prompts offer high-level contextual understanding, while the low-resolution images introduce spatial and structural constraints. These constraints are injected into the Stable Diffusion model in a side-controlling manner. By jointly utilizing these two types of control signals, our approach ensures that the generation of high-resolution images considers both semantic context and fine-grained textures required in medical imaging.

Our strategy comprises three key components: a text encoder for prompt encoding, a visual encoder that encodes low-resolution images, and a condition network that incorporates the encoded features into the diffusion model to guide the generation process.

Text Encoder. We leverage the text encoder of CLIP [28] to extract linguistic features, as it is pre-trained on massive image-text pairs. We freeze the text encoder to preserve its robust generalization in image-text alignment. Let $\mathcal{E}_{\text{T}}$ denote the text encoder, the encoding process can be written as:

$$f_p = \mathcal{E}_{\text{T}}(p). \tag{2}$$

Visual Encoder. The visual encoder extracts visual features from low-resolution inputs to guide the generative process. We utilize the pre-trained VAE encoder of Stable Diffusion, denoted as $\mathcal{E}_{\text{V}}$, to project x_{lr} into the VAE's latent space:

$$f_{lr} = \mathcal{E}_{\text{V}}(x_{lr}). \tag{3}$$

Condition Network. The latent representation of a high-resolution image is obtained by the pre-trained VAE encoder, denoted z_0. Both diffusion and denoising processes occur in the latent space. During diffusion, Gaussian noise with variance $\beta_t \in (0,1)$ at time t is added to z_0 to produce the noisy latent:

$$z_t = \sqrt{\bar{\alpha}_t}\, z_0 + \sqrt{1-\bar{\alpha}_t}\, \epsilon, \tag{4}$$

where $\epsilon \sim \mathcal{N}(0, \mathbf{I})$, $\alpha_t = 1 - \beta_t$, and $\bar{\alpha}_t = \prod_{s=1}^{t} \alpha_s$. At sufficiently large t, the latent z_t approximates a standard Gaussian distribution.

We define the entire U-Net in the Stable Diffusion model as $\mathcal{F}_{\text{U-Net}}$. In the denoising process, we clone a trainable copy of the U-Net encoder $\mathcal{E}_{\text{U}}$ as our condition network (denoted as $\mathcal{F}_{\text{COND}}$), which processes condition information and outputs control signals. This copy strategy provides a good weight initialization for the condition network. Then, the concatenation of the condition $\boldsymbol{f}_{lr}$ and the noisy latent z_t at time t forms z'_t. This concatenated representation, along with the textual embedding $\boldsymbol{f}_p$, are fed into $\mathcal{F}_{\text{COND}}$. At each block of $\mathcal{F}_{\text{COND}}$, the output is skip-connected to the corresponding block of the U-Net decoder $\mathcal{D}_{\text{U}}$ in Stable Diffusion, facilitating the effective integration of the conditional signals.

Meanwhile, $\boldsymbol{f}_p$ and z_t are fed into $\mathcal{F}_{\text{U-Net}}$. We denote the output of the i-th block of $\mathcal{D}_{\text{U}}$ as $\boldsymbol{o}_i$. In the i-th block where control information is involved, the calculation proceeds as follows:

$$\boldsymbol{o}_i = \mathcal{D}_{\text{U}}^i(\boldsymbol{o}_{i-1}, \boldsymbol{f}_p) + \mathcal{F}_{\text{COND}}^i(z'_t, \boldsymbol{f}_p). \tag{5}$$

The final high-resolution image is generated by mapping the denoised latent z'_0 through the VAE decoder $\mathcal{D}_{\text{V}}$:

$$\boldsymbol{x}'_{hr} = \mathcal{D}_{\text{V}}(z'_0). \tag{6}$$

2.3 Training Objective

We train a noise prediction model ϵ_θ conditioned on low-resolution images and text prompts:

$$\mathcal{L} = \mathbb{E}_{z_0, \boldsymbol{f}_{lr}, t, \boldsymbol{f}_p, \epsilon \sim \mathcal{N}} \left[\| \epsilon - \epsilon_\theta(z_t, t, \boldsymbol{f}_{lr}, \boldsymbol{f}_p) \|_2^2 \right]. \tag{7}$$

3 Experiments

3.1 Datasets and Evaluation Metrics

We evaluate our method on two widely used public CT image datasets: Pancreas [18] and 3D-IRCADb [18]. The Pancreas dataset contains 19,328 CT slices with a resolution of 512×512 pixels from 82 patients. For our experiments, we utilize 5,638 slices from 65 patients for training and 1,421 slices from 17 patients for testing. The 3D-IRCADb dataset comprises 2,823 CT slices (512×512 resolution) from 20 patients, with 1,663 slices from 16 patients for training and 411 slices from 4 patients for testing. Following [29], we standardize the data by clipping Hounsfield unit values to $[-135, 215]$ and simulate low-dose noise by setting the blank flux to 0.5×10^5. We adopt the degradation pipeline from Real-ESRGAN [12] to generate low-resolution images at 256×256 and 128×128 pixels, corresponding to upscaling factors of ×2 and ×4, respectively.

For evaluation, we employ two metrics: peak signal-to-noise ratio (PSNR) and structural similarity index measure (SSIM) [20]. They are widely accepted reference-based fidelity measures for assessing image quality.

3.2 Implementation Details

All experiments are conducted using PyTorch on NVIDIA RTX 4090Ti GPUs. We train the model for 150K iterations using the Adam optimizer [19] with a learning rate of 5×10^{-5} and a batch size of 2 per GPU across 8 GPUs. During inference, we employ spaced DDPM sampling [30] with 50 timesteps.

Table 1. Quantitative comparison with state-of-the-art methods on two public datasets at scale factors of ×2 and ×4. Results in **bold** and <u>underlined</u> indicate the best and second-best performance, respectively.

	Pancreas				3D-IRCADb			
	×2		×4		×2		×4	
	PSNR	SSIM	PSNR	SSIM	PSNR	SSIM	PSNR	SSIM
SRGAN	26.8732	0.7701	25.6810	0.7539	25.5262	0.7425	24.5693	0.7206
EDSR	28.8972	0.8364	27.8862	0.8105	28.2912	0.7973	27.1478	0.7882
ESRGAN	27.9114	0.8073	26.7801	0.7818	26.6362	0.7492	25.1963	0.7366
RCAN	29.1973	0.8442	28.0183	0.8236	28.5914	0.7969	27.2284	0.7891
IKC	29.7129	0.8709	28.4862	0.8449	29.4764	0.8254	28.2687	0.7914
DAN	<u>30.0293</u>	<u>0.8813</u>	<u>28.5829</u>	<u>0.8589</u>	<u>29.5708</u>	<u>0.8269</u>	<u>28.3362</u>	<u>0.7993</u>
RealSR	28.8903	0.8256	27.2931	0.8023	28.7308	0.8081	26.8275	0.7692
Real-ESRGAN	28.7841	0.8179	26.9721	0.7993	27.0132	0.7707	25.4315	0.7419
SwinIR	28.9764	0.8378	27.7532	0.8043	28.6215	0.8012	27.1922	0.7891
SR3	25.6895	0.7525	24.8917	0.7315	25.2515	0.7402	23.9709	0.7059
DFD-DCC	28.7357	0.8104	27.8532	0.8073	26.6801	0.7821	25.4221	0.7137
MDA-SR	28.7601	0.8647	27.0821	0.8290	27.7921	0.8082	26.8732	0.7671
RGT	28.9527	0.8755	27.8601	0.8575	29.5512	0.8253	28.3097	0.7899
Ours	**32.0349**	**0.8933**	**30.8082**	**0.8748**	**30.8707**	**0.8370**	**29.6483**	**0.8113**

3.3 Comparison with State-of-the-Art Methods

We compare our method against twelve state-of-the-art approaches: SRGAN [5], EDSR [7], ESRGAN [6], RCAN [8], IKC [10], DAN [11], RealSR [31], Real-ESRGAN [12], SwinIR [9], SR3 [13], DFD-DCC [14], MDA-SR [15], and RGT [16]. Table 1 demonstrates our framework's superior performance across both datasets and scaling factors. On the Pancreas dataset, our method surpasses the second-best approach by 2.0056dB/0.012 and 2.2253dB/0.0159 in PSNR and SSIM for ×2 and ×4 scaling factors, respectively. Similarly, for the 3D-IRCADb dataset, we achieve improvements of 1.2999dB/0.0101 and 1.3121dB/0.012 in PSNR and SSIM across both scales compared to the next best method.

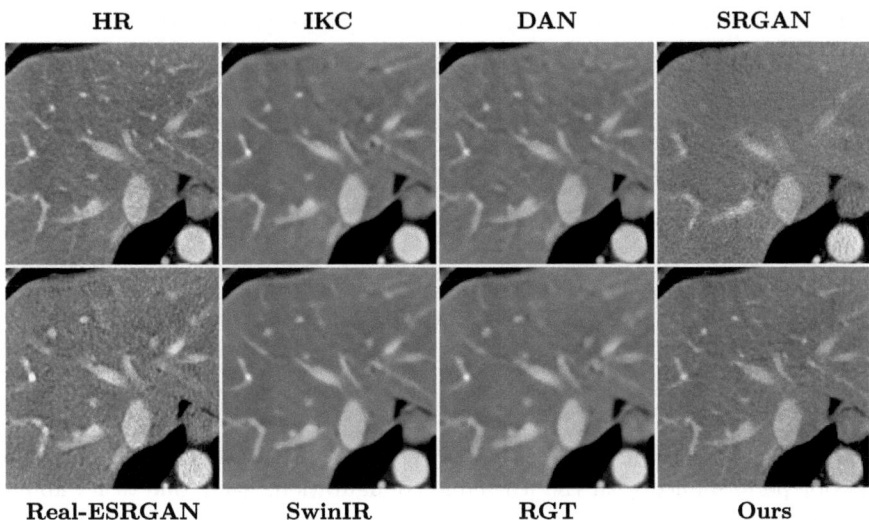

Fig. 2. Qualitative results on the 3D-IRCADb dataset with a scale factor of 4. The top-left image shows a zoomed-in region of a high-resolution image, focusing on the liver and its internal structures for better visualization. Subsequent images display super-resolution results from various methods. The bottom-right image presents our proposed method's output, with red arrows highlighting regions where our approach generates more accurate and detailed structures.

Figure 2 presents a qualitative comparison between our method and competing approaches. Our results demonstrate superior artifact suppression, particularly in edge regions and internal structures, while other methods suffer from artifacts and blurred structures. Furthermore, our approach produces semantically accurate and detail-rich reconstructions, preserving clear edges, fine details, and textural structures.

3.4 Ablation Studies

We conduct comprehensive ablation studies to validate each component of our framework.

Text Prompting Analysis. We evaluate the impact of text-based guidance by comparing our full model against variants with omitted text prompts or modified LLaVA-Med instructions. Results in Table 2 show that removing text prompts significantly degrades performance, highlighting their crucial role in providing valuable prior knowledge. Notably, our framework demonstrates robustness to minor variations in LLaVA-Med instructions, maintaining consistent performance—a crucial feature for practical applications.

Table 2. Ablation study on the Pancreas dataset comparing different instructions to LLaVA-Med. ✗: no text prompts; ♠: using instruction *Describe the anatomical structures in this CT image of the abdomen*; ♦: using instruction *List the major anatomic structures in this CT image of the abdomen*.

Instruction	×2		×4	
	PSNR	SSIM	PSNR	SSIM
✗	29.9012	0.8728	28.3015	0.8505
♠	32.0163	0.8928	30.7891	0.8724
♦	**32.0349**	**0.8933**	**30.8082**	**0.8748**

Visual Encoder. We examine two configurations of the pre-trained VAE encoder from Stable Diffusion: learnable versus frozen. Results in Fig. 3 indicate superior performance with the learnable configuration, likely due to its ability to adapt to medical image-specific features.

Condition Network. We investigate two initialization strategies for the condition network: pre-trained weights from Stable Diffusion's UNet versus random initialization. Figure 3 shows that pre-trained initialization yields better results in both PSNR and SSIM. This suggests that leveraging pre-trained weights enables more effective integration of conditional signals from low-resolution medical images and text prompts, ultimately leading to improved high-resolution image generation.

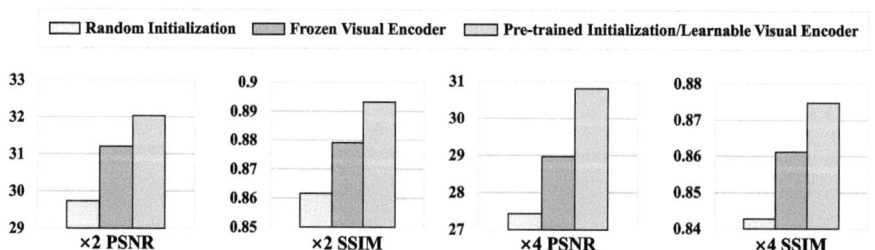

Fig. 3. Ablation study results on the Pancreas dataset. We evaluate the following configurations: random initialization (yellow) vs. pre-trained initialization (green), and frozen visual encoder (pink) vs. learnable visual encoder (green), for both ×2 and ×4 upscaling factors. Performance is measured using PSNR and SSIM. Our method consistently achieves the best results across all experimental settings. (Color figure online)

4 Conclusion and Future Work

In this work, we present a novel framework for CT blind super-resolution that successfully adapts Stable Diffusion to the medical imaging domain. By using LLaVA-Med for semantic description generation and introducing a side-controlling strategy, we effectively bridge the domain gap between natural and medical images. Our experimental results validate the framework's effectiveness, showing consistent improvements over existing methods.

The success of our approach suggests that pre-trained generative models can be effectively adapted for specialized medical imaging tasks. Looking ahead, the framework could be extended to other imaging modalities, and the integration of more sophisticated anatomical priors could further improve performance. We believe our work represents a step toward achieving high-quality CT imaging at reduced radiation doses, with potential impact on patient care through safer diagnostic procedures.

Future work could explore extending this framework to other medical imaging modalities such as MRI and ultrasound to further validate its generalizability in various healthcare applications.

References

1. Chu, Y., Zhou, L., Luo, G., Qiu, Z., Gao, X.: Topology-preserving computed tomography super-resolution based on dual-stream diffusion model. In: Medical Image Computing and Computer Assisted Intervention, pp. 260–270. (2023)
2. Akagi, M., et al.: Deep learning reconstruction improves image quality of abdominal ultra-high-resolution CT. Eur. Radiol. , 1–9 (2019). https://doi.org/10.1007/s00330-019-06170-3
3. You, C., et al.: CT super-resolution GAN constrained by the identical, residual, and cycle learning ensemble (GAN-CIRCLE). IEEE Trans. Med. Imaging **39**(1), 188–203 (2019)
4. Chen, Y., Zheng, Q., Chen, J.: Double paths network with residual information distillation for improving lung CT image super resolution. Biomed. Signal Process. Control **73**, 103412 (2022)
5. Ledig, C., et al.: Photo-realistic single image super-resolution using a generative adversarial network. In: IEEE Conference on Computer Vision and Pattern Recognition, pp. 4681–4690. (2017)
6. Wang, X., et al.: ESRGAN: Enhanced super-resolution generative adversarial networks. In: European Conference on Computer Vision Workshops, pp. 63–79. (2018)
7. Lim, B., Son, S., Kim, H., Nah, S., Mu Lee, K.: Enhanced deep residual networks for single image super-resolution. In: IEEE Conference on Computer Vision and Pattern Recognition Workshops, pp. 136–144 (2017)
8. Zhang, Y., Li, K., Li, K., Wang, L., Zhong, B., Fu, Y.: Image super-resolution using very deep residual channel attention networks. In: European Conference on Computer Vision, pp. 286–301 (2018)
9. Liang, J., Cao, J., Sun, G., Zhang, K., Van Gool, L., Timofte, R.: SwinIR: image restoration using Swin Transformer. In: Proceedings of the IEEE/CVF International Conference on Computer Vision Workshops, pp. 1833–1844 (2021)

10. Gu, J., Lu, H., Zuo, W., Dong, C.: Blind super-resolution with iterative kernel correction. In: Proceedings of the IEEE/CVF Conference on Computer Vision and Pattern Recognition, pp. 1604–1613 (2019)
11. Huang, Y., Li, S., Wang, L., Tan, T.: Unfolding the alternating optimization for blind super resolution. In: Advances in Neural Information Processing Systems, pp. 5632–5643. (2020)
12. Wang, X., Xie, L., Dong, C., Shan, Y.: Real-ESRGAN: training real-world blind super-resolution with pure synthetic data. In: Proceedings of the IEEE/CVF International Conference on Computer Vision Workshops, pp. 1905–1914 (2021)
13. Saharia, C., Ho, J., Chan, W., Salimans, T., Fleet, D.J., Norouzi, M.: Image super-resolution via iterative refinement. IEEE Trans. Pattern Anal. Mach. Intell. **45**(4), 4713–4726 (2022)
14. Chi, J., Sun, Z., Zhao, T., Wang, H., Yu, X., Wu, C.: Low-dose CT image super-resolution network with dual-guidance feature distillation and dual-path content communication. In: Medical Image Computing and Computer Assisted Intervention, pp. 98–108 (2023)
15. Liu, T., et al.: MDA-SR: Multi-level domain adaptation super-resolution for wireless capsule endoscopy images. In: Medical Image Computing and Computer Assisted Intervention, pp. 518–527 (2023)
16. Chen, Z., Zhang, Y., Gu, J., Kong, L., Yang, X.: Recursive generalization transformer for image super-resolution. In: International Conference on Learning Representations, (2023)
17. Rombach, R., Blattmann, A., Lorenz, D., Esser, P., Ommer, B.: High-resolution image synthesis with latent diffusion models. In: Proceedings of the IEEE/CVF Conference on Computer Vision and Pattern Recognition, pp. 10684–10695 (2019)
18. Clark, K.W., et al.: The cancer imaging archive (TCIA): maintaining and operating a public information repository. J. Digit. Imaging **26**(6), 1045–1057 (2013)
19. Kingma, D. P., Ba, J.: Adam: a method for stochastic optimization. arXiv preprint arXiv:1412.6980 (2014)
20. Zhou, W., Alan, C.B., Hamid, R.S., Eero, P.S.: Image quality assessment: from error visibility to structural similarity. IEEE Trans. Image Process. **13**(4), 600–612 (2004)
21. Li, C., et al.: LLaVA-Med: training a large language-and-vision assistant for biomedicine in one day. In: Advances in Neural Information Processing Systems, pp. 28541–28564 (2023)
22. Huang, Y., Wang, Q., Omachi, S.: Rethinking degradation: radiograph super-resolution via aid-srgan. In: International Workshop on Machine Learning in Medical Imaging, pp. 43–52 (2022)
23. Zhou, H., Huang, Y., Li, Y., Zhou, Y., Zheng, Y.: Blind super-resolution of 3D MRI via unsupervised domain transformation. IEEE J. Biomed. Health Inform. **27**(3), 1409–1418 (2022)
24. Lin, T., Chen, Z., Yan, Z., Yu, W., Zheng, F.: Stable diffusion segmentation for biomedical images with single-step reverse process. In: Medical Image Computing and Computer Assisted Intervention, pp. 656–666 (2024)
25. Tian, J., Aggarwal, L., Colaco, A., Kira, Z., Gonzalez-Franco, M.: Diffuse attend and segment: unsupervised zero-shot segmentation using Stable Diffusion. In: Proceedings of the IEEE/CVF Conference on Computer Vision and Pattern Recognition, pp. 3554–3563 (2024)
26. Cho, J., Aghajanzadeh, S., Zhu, Z., Forsyth, D. A.: Zero-shot low light image enhancement with diffusion prior. arXiv preprint arXiv:2412.13401 (2024)

27. Jiang, H., Luo, A., Fan, H., Han, S., Liu, S.: Low-light image enhancement with wavelet-based diffusion models. ACM Trans. Graph. **42**(6), 1–14 (2023)
28. Radford, A., et al.: Learning transferable visual models from natural language supervision. In: International Conference on Machine Learning, pp. 8748–8763 (2021)
29. Zeng, D., et al.: A simple low-dose X-ray CT simulation from high-dose scan. IEEE Trans. Nucl. Sci. **62**(5), 2226–2233 (2015)
30. Alexander Q.N., Prafulla D.: Improved denoising diffusion probabilistic models. In: International Conference on Machine Learning, pp. 8162–8171 (2021)
31. Ji, X., Cao, Y., Tai, Y., Wang, C., Li, J., Huang, F.: Real-world super-resolution via kernel estimation and noise injection. In: Proceedings of the IEEE/CVF Conference on Computer Vision and Pattern Recognition Workshops, pp. 466–467 (2020)

RadiSimCLIP: A Radiology Vision-Language Model Pretrained on Simulated Radiologist Learning Dataset for Zero-Shot Medical Image Understanding

Minhui Tan[1,2], Qingxia Wu[3,4], Boyang Zhang[2], Genqiang Ren[2], Jianlong Nie[1,2], Zhong Xue[1,2], Xiaohuan Cao[1,2(✉)], and Dinggang Shen[1,2,5(✉)]

[1] School of Biomedical Engineering & State Key Laboratory of Advanced Medical Materials and Devices, ShanghaiTech University, Shanghai 201210, China
xiaohuan.cao@uii-ai.com, Dinggang.Shen@gmail.com
[2] Shanghai United Imaging Intelligence Co., Ltd., Shanghai 200232, China
[3] Beijing United Imaging Research Institute of Intelligent Imaging, Beijing 100089, China
[4] United Imaging Intelligence (Beijing) Co., Ltd, Beijing 100089, China
[5] Shanghai Clinical Research and Trial Center, Shanghai 201210, China

Abstract. Medical vision-language models (MVLMs) have shown significant promise in medical diagnosis. However, most existing approaches are constrained by limited modality coverage and often rely on datasets lacking real-world radiological diversity and semantic precision. Moreover, current models exhibit insufficient evaluation on fine-grained tasks. In this paper, we introduce **RadiSim**, a large-scale **Radi**ology dataset comprising 10.6 million image-text pairs from CT, MRI, and DR modalities, designed to **Sim**ulate the learning resources during radiologist training. Building on RadiSim, we propose **RadiSimCLIP**, a radiology-specific MVLM pretrained using a CLIP-style contrastive framework. The model progressively acquires multilevel capabilities—ranging from anatomical recognition and pixel-level interpretability to vision-language alignment—mirroring how radiologists gain expertise. To systematically evaluate generalizability, we further propose a three-part evaluation framework: (1) zero-shot classification, which includes a novel extension to 3D volumetric data, evaluating anatomical, modality, and disease recognition; (2) zero-shot organ detection, which evaluates pixel-level semantic localization without parameter tuning; and (3) cross-modal retrieval, measuring alignment between visual features and clinical text. RadiSimCLIP achieves state-of-the-art results on 15 of 16 downstream benchmarks. This work underscores the feasibility and potential of large-scale radiology-specific MVLMs for robust, zero-shot 3D medical understanding, representing a practical step toward clinically applicable foundation models. The GitHub repository is https://github.com/so-ux/RadiSimCLIP.

M. Tan and Q. Wu—These authors contributed equally to this work.

Keywords: vision-language pre-training model · zero-shot classification · zero-shot detection · cross-modal retrieval

1 Introduction

Radiological imaging is indispensable in modern medicine; however, the increasing volume and complexity of scans have placed significant strain on radiologists, resulting in heightened workloads, interpretation delays, and inconsistencies. This situation underscores the pressing demand for robust automated analysis systems.

In recent years, medical vision-language models (MVLMs) have emerged as promising tools to address these challenges by integrating complementary features that enhance the efficiency and precision of medical services [2]. These models have demonstrated remarkable capabilities in various medical imaging tasks, including detecting chest pathologies from X-rays [19,20,23,25,30], diagnosing gastrointestinal diseases via endoscopic videos [22], assessing cardiac function using cardiac ultrasound videos [15], and analyzing histopathology images [7,8,13,14].

Despite these advances, current models primarily focus on specific diseases and modalities and still face significant limitations when applied to other modalities of radiological imaging. Although some existing methods can be adapted to general medical data, they still come with certain drawbacks. For example, PMC-CLIP [12] and BiomedCLIP [28,29] are based primarily on PubMed Central data (PMC), which do not adequately represent the complexities of medical imaging in the real world [10], limiting their sensitivity to specialized radiological terminology and visual characteristics. Additionally, while these models have been validated for image-text retrieval and 2D image classification, they are rarely extended to 3D volumetric applications. Other approaches [3,6,11,27] have expanded their datasets, but often lack validation for tasks requiring fine-grained recognition and interpretability, such as zero-shot detection. This gap hinders their effectiveness in high-precision medical tasks, where the ability to locate lesions or organs without additional fine-tuning is crucial. However, few studies have explored the capabilities of CLIP-based models for zero-shot detection in a medical context.

To address these challenges, we first introduce RadiSim, a large-scale, multimodal radiology dataset comprising 10.6 million meticulously curated image-text pairs from CT, MRI, and DR modalities. The name "RadiSim" combines **Radi**ology and **Sim**ulating training resources—designed to emulate the diverse materials encountered during radiologist education, such as anatomical references and pathological case descriptions. Through large-scale pretraining on RadiSim using contrastive learning (e.g., CLIP), we propose RadiSimCLIP, which progressively acquires hierarchical competencies—from basic anatomical recognition to diagnostic reasoning—mimicking the structured knowledge acquisition process in clinical training. We establish a multi-granular evaluation protocol: 1) Zero-shot classification (2D/3D) for organ/modality/disease recognition,

2) Pixel-level organ detection without fine-tuning, and 3) Cross-modal retrieval for semantic alignment verification. RadiSimCLIP achieves state-of-the-art performance on 15 out of 16 radiological downstream tasks, demonstrating its strong generalization ability and potential as a universal radiology assistant.

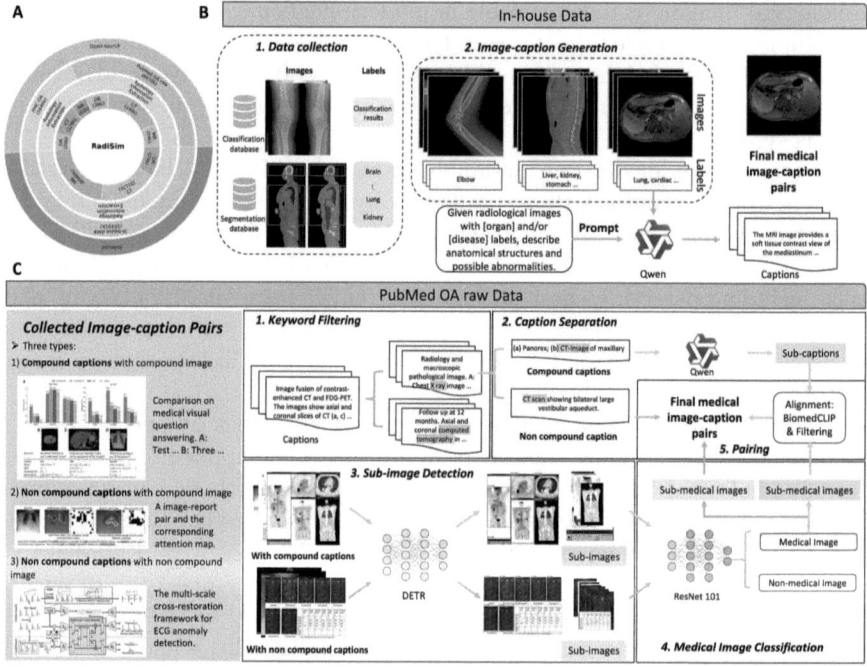

Fig. 1. Construction of the RadiSim dataset. (A) Overview of RadiSim, encompassing both in-house and publicly available data sources; (B) Workflow for constructing the in-house dataset; (C) Data curation pipeline for the PubMed OA raw dataset.

2 Construction of the RadiSim Dataset

To address the lack of representativeness and diversity in existing datasets, we constructed a comprehensive image-caption pair dataset named RadiSim. This dataset incorporates 10,585k in-house samples and 484k curated samples from diverse open-source datasets sourced from two public resources: PubMed Central (PubMed OA raw) [16] and PMC-OA [12] (Fig. 1A).

The in-house portion (Fig. 1B) is built via a two-stage pipeline: data curation and caption generation. We first collect a diverse set of images and associated disease or organ labels from internal classification and segmentation tasks across partner hospitals. While segmentation samples are limited, they help identify regions of interest, enhancing anatomical precision. Next, a domain-adapted large language model (Qwen2 [21]) is prompted with both image content and structured labels to produce detailed, anatomically grounded captions. This results in rich, aligned image-text pairs with high semantic fidelity. For PubMed OA raw, our data curation process (Fig. 1C), inspired by PMC-CLIP, categorizes content into three types: compound captions with compound images, non-compound captions with compound images, and non-compound captions with non-compound images. These types undergo several steps: 1) keyword filtering to select relevant radiology image-caption pairs; 2) caption separation using Qwen2 to decompose compound captions into sub-captions; 3) sub-image detection and cropping for compound images, 4) ResNet101 is employed to filter out non-medical images; and 5) consolidation of all qualifying pairs into the dataset, resulting in 202k radiological image-text pairs. For the PMC-OA dataset, we filtered out non-radiological pairs, retaining 232k relevant image-text pairs.

3 Experiments

3.1 RadiSimCLIP Training

We fine-tuned the pre-trained BiomedCLIP model on our RadiSim dataset to enhance its adaptation to radiological imaging (Fig. 2). Training followed a two-stage strategy to leverage data heterogeneity: we first warmed up the model on the external public subset to encourage general cross-modal alignment across diverse imaging styles, then fine-tuned it on the in-house dataset containing more clinically grounded semantics. The in-house data includes classification-derived pairs with anatomical and diagnostic labels, and segmentation-derived samples emphasizing organ-level spatial cues.

We adopted a selective freezing strategy, keeping most of the text encoder frozen except for LayerNorm parameters, while fully training the visual encoder and projection layers. The model was trained using AdamW with a batch size of 4096 for 30 epochs. Training was conducted on 8 NVIDIA L40 GPUs and took approximately one week to complete.

3.2 Downstream Datasets

To comprehensively evaluate the performance of RadiSimCLIP in anatomical understanding, fine-grained recognition, and general medical reasoning, we designed and conducted four downstream experiments across diverse datasets, as detailed in Table 1.

3.3 Downstream Tasks

2D Zero-Shot Classification. We first assess RadiSimCLIP's zero-shot capabilities on 2D medical images. For CT imaging, we conduct 2D slice-based eight-class classification across three anatomical planes (axial, sagittal, and coronal),

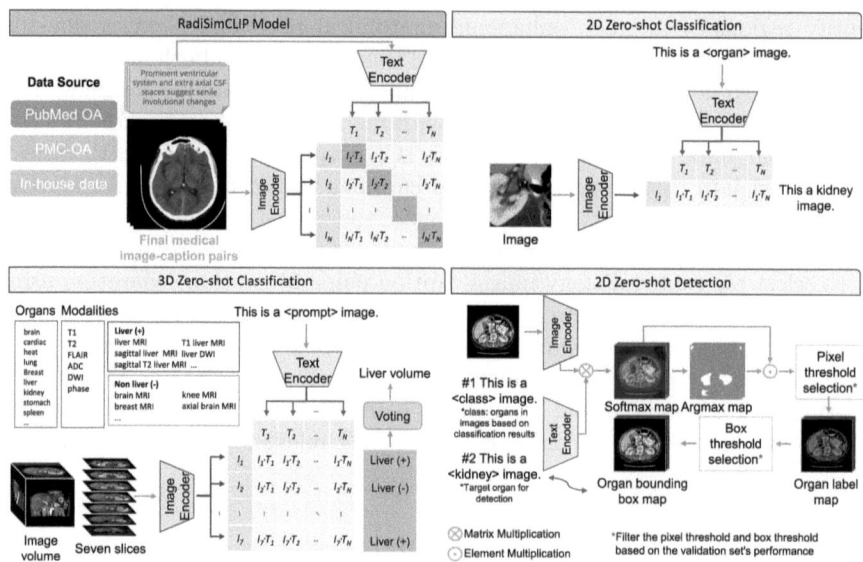

Fig. 2. Overview of RadiSimCLIP training and three typical downstream tasks.

while for DR imaging, we evaluate on PneumoniaMNIST [26] and RSNA Pneumonia [17] datasets. The model performs zero-shot classification by computing cosine similarity scores between image embeddings and pre-defined class-descriptive text embeddings, eliminating the need for fine-tuning. We compare our approach with state-of-the-art models including CLIP [18], MedCLIP [23], PubMedCLIP [4], PMC-CLIP, and BiomedCLIP.

3D Zero-Shot Classification. Our 3D volume zero-shot classification tasks are performed on the in-house dataset and open-source datasets (i.e., knee MRI ACL and Meniscus [1]), using a three-stage reasoning framework. Specifically, in the image embedding construction phase, we evaluated multiple 2D slice sampling strategies (e.g., median slice, maximum intensity slice, and uniform sampling) to balance computational efficiency and spatial coverage. Ultimately, we selected uniform sampling across seven slices along the z-axis to extract anatomical features. In the semantic constraint construction stage, we adhered to the T/CHIA 23-2021 standard classification codes for medical imaging examination sites to establish a constrained text generation paradigm. This involved creating a foundational template for organ-modal combinations (e.g., "The modality image shows [organ]") and employing a greedy algorithm to maximize the cosine similarity difference between positive and negative texts within the pool. During the decision-making fusion stage, we employed majority voting to aggregate the classification results of the volumetric data across N slices (in this work, N = 7).

Table 1. Overview of downstream tasks and evaluation datasets.

Task	Dataset	Metric	Description	Scale
2D Zero-shot Classification	CT-Axial	ACC	Eight-class organ classification (MediMeTA [24]), with bilateral organs merged	1,645 images
	CT-Sagittal	ACC		1,645 images
	CT-Coronal	ACC		1,645 images
	Pneumonia-MNIST [26]	AUC	Pneumonia vs. normal in chest X-rays	5,856 images
	RSNA Pneumonia [17]	AUC	Pneumonia vs. normal in chest X-rays	23,286 images
3D Zero-shot Classification	MRI organ Brain	ACC	Brain vs. other organs	63,910 volumes
	MRI organ Breast	ACC	Breast vs. other organs	63,910 volumes
	MR modality (TOF-MRA)	ACC	MR angiography (MRA) vs. conventional MRI	5,688 volumes
	MR modality (T1C)	ACC	T1 contrast vs. non-contrast classification	5,076 volumes
	Knee MRI disease (ACL) [1]	AUC	ACL tear vs. non-ACL tear using sagittal view	1,250 patients
	Knee MRI disease (Meniscus) [1]	AUC	Meniscal tear vs. nonmeniscal tear using sagittal view	1,250 patients
2D Zero-shot Detection	In-house Dataset	F1	Nine organ detection in MR volumes	1,117 detection boxes
	AMOS-CT [9]	F1	Seven organ detection in CT volumes	8,289 detection boxes
Cross-modal Retrieval	PubMed OA-MR	Top-1%ACC, Avg. Rank	Text-to-image retrieval and image-to-text retrieval	26,589 images
	PubMed OA-CT			18,008 images
	PubMed OA-DR			4,503 images

2D Zero-Shot Detection. To evaluate the accuracy of text-guided visual models in identifying fine-grained information, we constructed a zero-shot detection task framework (Fig. 2). Inspired by attention-based interpretation method [5], this framework decodes spatial attention patterns from the CLIP image encoder to localize anatomical structures. Specifically, we leverage text prompts describing organs to generate multi-label softmax attention maps through 12 attention heads in the final layer of the image encoder. These multi-head attention maps are aggregated to highlight regions semantically aligned with the input text. Using a predefined organ vocabulary, we then visualize the corresponding activation maps and apply a thresholding strategy to suppress low-confidence responses. Finally, bounding boxes are derived from the thresholded activation maps to localize target organs in the image.

During the experimental phase, we employed this framework for anatomical localization of 13 representative organs and a background class (i.e., brain, stomach, lung, heart, breast, liver, spleen, kidney, pancreas, prostate, bladder, uterus, knee, and others). Text embeddings for these 14 categories were generated via a pre-trained text encoder, then projected into a shared embedding space with image embeddings. Relative confidence scores were computed by normalizing cosine similarity, followed by softmax activation. To obtain accurate organ localization, we applied the threshold corresponding to the best F1 score calculated on the validation data set to generate a binary activation map on the test set. Subsequently, connected component analysis was performed on the thresholded maps to extract bounding boxes for specified organs.

Cross-Modal Retrieval. We utilized untrained data from PubMed-OA to assess the degree of the model correlating medical images with their corresponding textual descriptions and vice versa.

4 Results

4.1 2D Zero-Shot Classification

Table 2 presents a comprehensive comparison of 2D zero-shot classification performance across different models on medical imaging tasks. These results collectively demonstrate that RadiSimCLIP excels in both modality-specific recognition and lesion detection, outperforming other baseline models by significant margins. Specifically, RadiSimCLIP achieves remarkable performance on CT imaging across all three anatomical planes, attaining accuracy scores of 0.765, 0.683, and 0.711 for axial, sagittal, and coronal views, respectively. In DR tasks, RadiSimCLIP exhibits state-of-the-art performance on PneumoniaMNIST with an accuracy of 0.728, while BiomedCLIP slightly outperforms on the RSNA dataset with an accuracy of 0.765. In contrast, the general-purpose CLIP model shows limited effectiveness, achieving only 0.106 accuracy on CT tasks and random-chance performance (0.500 AUC) on DR tasks.

Table 2. Performances of zero-shot classification on different modalities and tasks.

Model	CT			DR		MRI organ		MR modality		MRI disease	
	Axial	Sag.	Cor.	MNIST	RSNA	Brain	Breast	MRA	T1C	ACL	Meniscus
CLIP	0.106	0.106	0.106	0.500	0.500	0.787	0.716	0.155	0.908	0.620	0.582
MedCLIP	0.261	0.292	0.314	0.498	0.711	0.763	0.888	0.270	0.852	0.534	0.551
PubMedCLIP	0.419	0.343	0.398	0.553	0.646	0.999	0.944	0.864	0.949	0.445	0.509
PMC-CLIP	0.300	0.272	0.277	0.556	0.579	0.909	0.803	0.647	0.937	0.618	0.427
BiomedCLIP	0.582	0.617	0.644	0.689	**0.765**	1.000	0.998	**1.000**	0.998	0.451	0.430
RadiSimCLIP	**0.765**	**0.683**	**0.711**	**0.728**	0.761	**1.000**	0.999	**1.000**	**1.000**	**0.707**	**0.643**

4.2 3D Zero-Shot Classification

We validated the classification capability of RadiSimCLIP for organs, modalities, and diseases in MR data, with the results presented in Table 2. In the organ binary classification task, RadiSimCLIP achieved outstanding performance, attaining an accuracy close to 1.000. Regarding modality classification, RadiSimCLIP also exhibited high accuracy. In the disease multi-classification task, RadiSimCLIP achieved the highest score in the AUC metric, with scores of 0.707 for ACL tears and 0.643 for meniscus tears, significantly outperforming other models. These results suggest that RadiSimCLIP possesses exceptional capabilities in identifying disease-related patterns within zero-shot MRI data and demonstrates a superior understanding of image details compared to alternative models.

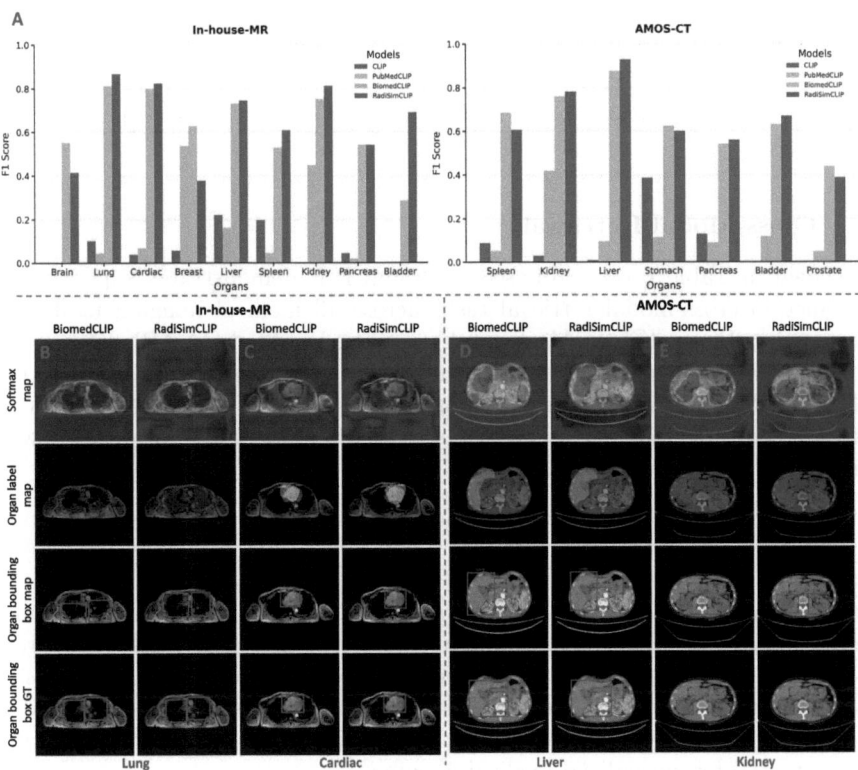

Fig. 3. Results of zero-shot detection across various organs in different datasets.

4.3 2D Zero-Shot Detection

We applied our zero-shot detection framework to CLIP, PubMedCLIP, BiomedCLIP, and RadiSimCLIP (Fig. 3). From Fig. 3A, RadiSimCLIP achieved the highest scores in seven out of nine organ categories on internal datasets, and showed superior performance on most organ categories in the AMOS-CT dataset. Due to the limited softmax map content in CLIP and PubMedCLIP, we focus our visual comparison on BiomedCLIP and our method. Our results demonstrate three key advantages: (1) more precise organ localization, with softmax maps concentrated on target organs, (2) improved alignment with organ morphology, closely matching ground-truth boundaries, and (3) accurate organ number detection through threshold-based feature screening, beneficial particularly for paired organs such as the lungs and kidneys.

Table 3. Performance of cross-modal retrieval.

Dataset	BiomedCLIP		RadiSimCLIP	
	Top1% Acc ↑	Avg. Rank ↓	Top1% Acc ↑	Avg. Rank ↓
PubMed OA-MR	0.428	2016	0.763	523
PubMed OA-CT	0.698	457	0.781	310
PubMed OA-DR	0.655	111	0.717	97

4.4 Cross-Modal Retrieval

As shown in Table 3, our proposed RadiSimCLIP demonstrates superior performance in cross-modal retrieval tasks across all medical imaging modalities compared to BiomedCLIP. Most notably, on the PubMed OA-MR dataset, our method achieves a substantial improvement with a Top1% accuracy of 0.763 (vs. 0.428) and reduces the average rank from 2016 to 523. Similar improvements are also observed in CT (0.781 vs. 0.698) and DR (0.717 vs. 0.655) modalities, with consistently lower average ranks.

5 Conclusion

We have constructed RadiSim, a radiology-specific dataset with 10.6M multi-source image-text pairs. Using RadiSim, RadiSimCLIP achieves SOTA accuracy on 15/16 tasks via hierarchical validation: 1) anatomical recognition via zero-shot classification, 2) pixel-level understanding through zero-shot detection, and 3) semantic alignment via cross-modal retrieval. This demonstrates general-purpose AI's potential for specialized medical imaging.

Acknowledgments. This work was supported in part by National Key Research and Development Program of China (No. 2022YFE0205700), National Natural Science Foundation of China (grant numbers 82441023, U23A20295, 62131015), Shanghai Municipal Central Guided Local Science and Technology Development Fund (No. YDZX20233100001001), Beijing Natural Science Foundation (IS24053), and HPC Platform of ShanghaiTech University and Shanghai United Imaging Intelligence Co., Ltd.

References

1. Bien, N., et al.: Deep-learning-assisted diagnosis for knee magnetic resonance imaging: development and retrospective validation of mrnet. PLoS Med. **15**(11), e1002699 (2018)
2. Chen, Q., et al.: A survey of medical vision-and-language applications and their techniques. arXiv preprint arXiv:2411.12195 (2024)
3. Codella, N.C., et al.: Medimageinsight: an open-source embedding model for general domain medical imaging. arXiv preprint arXiv:2410.06542 (2024)
4. Eslami, S., Meinel, C., De Melo, G.: Pubmedclip: How much does clip benefit visual question answering in the medical domain? In: Findings of the Association for Computational Linguistics: EACL 2023, pp. 1181–1193 (2023)
5. Gandelsman, Y., Efros, A.A., Steinhardt, J.: Interpreting clip's image representation via text-based decomposition. arXiv preprint arXiv:2310.05916 (2023)
6. He, X., et al.: Unified medical image pre-training in language-guided common semantic space. In: European Conference on Computer Vision, pp. 123–139. Springer (2025)
7. Huang, Z., Bianchi, F., Yuksekgonul, M., Montine, T.J., Zou, J.: A visual-language foundation model for pathology image analysis using medical twitter. Nat. Med. **29**(9), 2307–2316 (2023)
8. Ikezogwo, W., et al.: Quilt-1m: one million image-text pairs for histopathology. In: Advances in Neural Information Processing Systems, vol. 36 (2024)
9. Ji, Y., et al.: Amos: a large-scale abdominal multi-organ benchmark for versatile medical image segmentation (2022). https://arxiv.org/abs/2206.08023
10. Kakkar, M., Shanbhag, D., Aladahalli, C., Reddy, G.: Language augmentation in clip for improved anatomy detection on multi-modal medical images. In: 2024 46th Annual International Conference of the IEEE Engineering in Medicine and Biology Society (EMBC), pp. 1–4. IEEE (2024)
11. Khattak, M.U., Kunhimon, S., Naseer, M., Khan, S., Khan, F.S.: Unimed-clip: towards a unified image-text pretraining paradigm for diverse medical imaging modalities. arXiv preprint arXiv:2412.10372 (2024)
12. Lin, W., et al.: PMC-clip: contrastive language-image pre-training using biomedical documents. In: International Conference on Medical Image Computing and Computer-Assisted Intervention, pp. 525–536. Springer (2023)
13. Lu, M.Y., et al.: A visual-language foundation model for computational pathology. Nat. Med. **30**(3), 863–874 (2024)
14. Lu, M.Y., et al.: Visual language pretrained multiple instance zero-shot transfer for histopathology images. In: Proceedings of the IEEE/CVF Conference on Computer Vision and Pattern Recognition, pp. 19764–19775 (2023)
15. Vision-language foundation model for echocardiogram interpretation: Matthew Christensen, Milos Vukadinovic, N.Y., Ouyang, D. Nat. Med. **30**, 1481–1488 (2024)
16. National Library of Medicine: PMC open access subset (2003). https://pmc.ncbi.nlm.nih.gov/tools/openftlist/
17. of North America, R.S.: RSNA pneumonia detection challenge (2018). https://www.kaggle.com/c/rsna-pneumonia-detection-challenge/. Accessed: [date of access]
18. Radford, A., et al.: Learning transferable visual models from natural language supervision. In: International Conference on Machine Learning, pp. 8748–8763. PMLR (2021)

19. Rajpurkar, P.: Chexnet: Radiologist-level pneumonia detection on chest x-rays with deep learning. ArXiv abs/1711 **5225** (2017)
20. Tiu, E., Talius, E., Patel, P., Langlotz, C.P., Ng, A.Y., Rajpurkar, P.: Expert-level detection of pathologies from unannotated chest x-ray images via self-supervised learning. Nature Biomed. Eng. **6**(12), 1399–1406 (2022)
21. Wang, P., et al.: Qwen2-vl: enhancing vision-language model's perception of the world at any resolution. arXiv preprint arXiv:2409.12191 (2024)
22. Wang, Z., Liu, C., Zhang, S., Dou, Q.: Foundation model for endoscopy video analysis via large-scale self-supervised pre-train. In: International Conference on Medical Image Computing and Computer-Assisted Intervention, pp. 101–111. Springer (2023)
23. Wang, Z., Wu, Z., Agarwal, D., Sun, J.: Medclip: contrastive learning from unpaired medical images and text. arXiv preprint arXiv:2210.10163 (2022)
24. Woerner, S., Jaques, A., Baumgartner, C.F.: A comprehensive and easy-to-use multi-domain multi-task medical imaging meta-dataset (medimeta) (2024). https://arxiv.org/abs/2404.16000
25. Yang, H., et al.: Multi-modal vision-language model for generalizable annotation-free pathological lesions localization and clinical diagnosis. http://arxiv.org/abs/2401.02044
26. Yang, J., et al.: Medmnist v2-a large-scale lightweight benchmark for 2D and 3D biomedical image classification. Sci. Data **10**(1), 41 (2023)
27. Zhang, K., et al.: A generalist vision–language foundation model for diverse biomedical tasks. Nature Medi. 1–13 (2024)
28. Zhang, S., et al.: Large-scale domain-specific pretraining for biomedical vision-language processing. arXiv preprint arXiv:2303.00915 **2**(3), 6 (2023)
29. Zhang, S., et al.: Biomedclip: a multimodal biomedical foundation model pretrained from fifteen million scientific image-text pairs. arXiv preprint arXiv:2303.00915 (2023)
30. Zhang, X., Wu, C., Zhang, Y., Xie, W., Wang, Y.: Knowledge-enhanced visual-language pre-training on chest radiology images. Nat. Commun. **14**(1), 4542 (2023)

Improving Medical Visual Instruction Tuning with Labeled Datasets

Amin Dada[1], Amanda Butler Contreras[2], Constantin Seibold[1],
Osman Alperen Koraş[1], Julius Keyl[1,3], Aokun Chen[4], Cheng Peng[4],
Alexander Brehmer[1], Kaleb E. Smith[2], Jiang Bian[5,6], Yonghui Wu[4],
and Jens Kleesiek[1,7,8,9]

[1] Institute for AI in Medicine (IKIM), University Hospital Essen (AöR), Essen, Germany
amin.dada@uk-essen.de
[2] NVIDIA, Santa Clara, CA, USA
[3] Institute of Pathology, University Hospital Essen (AöR), Essen, Germany
[4] Department of Health Outcomes and Biomedical Informatics, College of Medicine, University of Florida, Gainesville, FL, USA
[5] Biostatistics and Health Data Science, School of Medicine, Indiana University, Indianapolis, IN, USA
[6] Regenstrief Institute, Indianapolis, IN, USA
[7] Cancer Research Center Cologne Essen (CCCE), West German Cancer Center Essen University Hospital Essen (AöR), Essen, Germany
[8] German Cancer Consortium (DKTK, Partner site Essen), Heidelberg, Germany
[9] Department of Physics, TU Dortmund, Dortmund, Germany

Abstract. We introduce the Medical Vision Instruction-tuning Dataset (MVID), a large-scale dataset containing 558,052 samples designed to enhance the instruction-following capabilities of medical vision-language models (VLMs). Previous vision-language datasets did not include clinical data as they were generated by commercial large language models (LLMs) that did not meet the privacy and regulatory constraints. Additionally, prior instruction-tuning efforts for medical VLMs focused on figure-caption pairs from PubMed, covering a broad array of medical topics but lacking clinical applications related themes, such as knowledge, detection, and localization of common findings. To address these limitations, MVID is generated with an open LLM and incorporates a diverse range of labeled data, including classification, segmentation, and free-text tasks, equipping models with essential medical competencies. Our experiments show that medical VLMs trained on MVID outperform those trained on PubMed-based datasets on VQA benchmarks by up to 20.29% in a zero-shot setting. Furthermore, our 7 billion parameter model demonstrates stronger performance than GPT-4o on radiology tasks, highlighting the effectiveness of MVID in developing capable medical VLMs.

Keywords: Visual Instructions · Multimodal Large Language Model · Medical Visual Question Answering

1 Introduction

Medical Vision Language Models (VLMs) have the potential to serve as generalist assistants in clinical workflow due to their zero-shot abilities [27]. While proprietary medical VLMs seem promising [34], the effectiveness of open-source alternatives remains questionable [16]. However, specifically in the clinical domain, open-source models are usually preferred due to privacy regulations. Open-source approaches often rely on the reformulation of PubMed figure captions to visual instructions using GPT models [24]. This poses several issues for clinical use-cases: (1) PubMed articles include a broad range of biomedical topics, which do not necessarily align with common clinical situations. (2) Precision is crucial for clinical applications, but the automatic linking of PubMed figures and captions is noisy [6] (3) GPT cannot be applied to private data.

A practical solution for quality issues is creating visual instructions leveraging existing medical imaging datasets to which a plethora of expert-created classification labels, segmentation masks, bounding boxes, and imaging reports are attached. Additionally, advances in open-weight LLMs have reduced the gap between proprietary models [8], offering an alternative to GPT-based visual instruction generation.

In this paper we utilize large collections of medical imaging data annotated by experts and specialized models. We introduce a methodology that repurposes these high-quality annotations to generate reliable visual instructions for medical VLM training. Additionally, we utilize a locally deployed LLM, ensuring that our method remains applicable to private clinical datasets while maintaining high standards of data security and compliance.

Our contributions are threefold: (1) We generate and publicly release MVID[1], a dataset comprising **558,052 visual instruction samples** derived from labeled medical imaging datasets using an open-weight LLM; (2) We train medical VLMs on MVID and demonstrate **up to 20.29% higher average accuracy** compared to models trained on PubMed-derived instructions across multiple medical visual question answering (VQA) benchmarks; (3) We introduce and release[2] **new medical VLMs with 7 billion parameters that perform competitively with GPT-4o**, advancing the development of clinically applicable, privacy-compliant, and open-source medical VLMs.

2 Related Work

The LLaVA approach [26] has been shown to produce well-performing models on various benchmarks using a simplistic architecture: A two-layer perceptron that maps image embeddings to the text embedding space. LLaVA relies on visual instruction tuning data generated by GPT-4, which is derived from existing image captions. Following its success, this approach was extended to

---

[1] https://huggingface.co/datasets/ikim-uk-essen/MVID.
[2] https://huggingface.co/ikim-uk-essen/MVID-Qwen2.5-7B.

the medical domain by applying the same methodology to PubMed figure captions, leading to the development of Llava-Med [24]. More recently, improvements have been made by querying larger commercial VLMs to refine the generated visual instruction tuning data [35] or to create new visual instruction tuning datasets [6]. However, these enhancements can only be applied to non-private data, limiting their applicability to clinical datasets.

3 Medical Visual Instruction Dataset

We compiled MVID, a dataset of 558,052 medical visual instructions collected from 29 medical imaging datasets spanning nine medical domains and five types of annotations, as detailed in Table 1. We apply different methods to construct textual descriptions of the given annotations. Based on these descriptions, we generate conversations between physicians and a VLM using the open-weight LLM *Llama-3.3-70B-Instruct* [8]. An overview is illustrated in Fig. 1. The following two sections describe the details of the generation process.

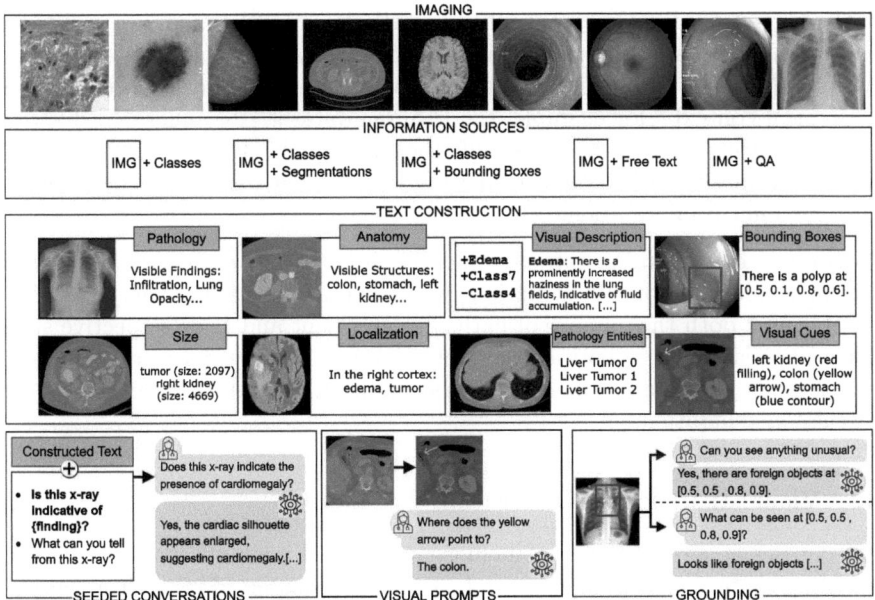

Fig. 1. Data generation process of MVID. We include data from 9 different domains with varying types of labels serving as information sources. We then derive enriched textual descriptions by applying multiple methods. These descriptions are then fed to an LLM to generate visual instructions with prompting methods.

3.1 Construction of Textual Descriptions

For all classification, segmentation and bounding box datasets we list the given pathological and anatomical features. Some labels correspond to specific visual properties. For example, the presence of a pleural effusion is indicated by noticeable blunting of the costophrenic angles. Where applicable, we formulated **Visual Descriptions** of such labels to enhance the available information for generating visual instructions.

For CT and MRI datasets, we sample individual slices. For the pathology dataset, we ensure an equal distribution of slices with and without pathologies. In the case of Atlas [15], a CT dataset containing segmentations of 142 anatomical structures, we sample one slice for each unique combination of segmentation masks. We derived various forms of spatial information from segmentation masks:

Mask Sizes. We include the segmentation size in pixels in the textual descriptions. When generating visual instructions, we instruct the LLM to include this information without mentioning specific pixel sizes (e.g., how large a tumor is in relation to the affected organ).

Pathology Localization. To enhance pathology localization, we generated anatomical segmentation masks using TotalSegmentator [38] for CT images and SynthSeg [4] for brain MRIs. Pathology segmentation masks often lacked detailed anatomical context—for example, the RibFrac dataset [18] annotated different types of rib fractures but did not specify the affected rib. By combining generated anatomical masks with pathology masks, we enriched the textual descriptions with anatomical information.

Pathology Entities. We extracted individual pathology entities by computing the connected components of pathology segmentation masks. This allowed us to determine both the number of pathologies present and their respective sizes. When applicable, these entities were also mapped to anatomical regions using the methods described earlier, providing additional spatial context for visual instruction generation.

3.2 Visual Instruction Generation

Our dataset generation process includes specialized prompts for each dataset, providing contextual explanations and outlining the expected visual instructions. These prompts were refined iteratively by examining generated visual instruction samples. Additionally, we applied different prompting strategies to formulate diverse visual instructions from the constructed textual descriptions.

Conversation Seeds. While using a temperature of 1.0 allows for variability in generated conversations, we found that the outputs were often identical for the same inputs, especially, in classification datasets with a limited number of labels, leading to repetitive results. To address this, we use conversation seeds—a set of introductory statements in conversations. For classification datasets, these seeds included placeholders that were randomly replaced with a possible class label (e.g., "Does this X-ray show signs of {finding}?", where finding is replaced with a randomly selected class for each conversation). To ensure diversity, each generated conversation was forced to begin with a randomly selected conversation seed.

Visual Prompts. To enhance the visual instruction tuning dataset, we introduced visual prompts by incorporating segmentation masks with distinct **Visual Cues**. These cues were assigned randomly from a predefined color set and presented in various forms, such as arrows, contours, or filled regions. The visual prompts were overlaid onto the images, with their color and shape included in the corresponding generation prompts.

Groundings. To enable visual grounding in generated conversations, we incorporated datasets with **Bounding Box** annotations. Each bounding box was represented using the corresponding label followed by a unique identifier tag (e.g., Polyp<Bounding Box 1>). These tags were then included in the input information for visual instruction generation. The LLM was explicitly instructed to utilize these tags, either by integrating them into user queries or by referencing them in the VLM's responses. As a post-processing step, the grounding tags were replaced with the corresponding bounding box coordinates $[X_{min}, Y_{min}, X_{max}, Y_{max}]$, normalized to a range between $[0, 1]$.

4 Experiments

To assess the efficiency of MVID, we train multiple VLMs and compare their performance with existing models. Our experiments are structured into two parts:

Experiment 1: Direct Comparison with Llava-Med. We train a model on MVID using the same configuration as LLaVA-Med-v1.5 [24]: The LLaVA 1.5 architecture [26], a 336×336 CLIP-Large image encoder [33] and the Mistral-7B-Instruct-v0.2 LLM[3]. We compare this configuration to a 224 × 224 BioClip image encoder [40], the biomedical LLM Biomistral [22] and Qwen2.5-7B-Instruct [39]. After identifying the best configuration, we train a model on the LLaVA-Med dataset for further comparison. We select a random subset of 60,000 conversations from MVID to ensure a fair comparison to LLaVA-Med.

Experiment 2: Full MVID Training: Building on the insights from Experiment 1, we train a model on the entire MVID dataset and compare its performance against various existing VLMs.

4.1 Training

For a direct comparison with LLaVA-Med, we train all models on the LLaVA-Med alignment dataset using a learning rate of $2e^{-3}$, a cosine learning rate

---
[3] https://huggingface.co/mistralai/Mistral-7B-Instruct-v0.2.

Table 1. All labeled imaging datasets that were used for the generation of MVID. **#Conv.** refers to the number of visual instruction conversations that were generated for each dataset. The available labels include visual question-answer pairs (VQA), classification labels (Cls), segmentation masks (Seg.) and bounding boxes (BB). The augmentations are described in Sect. 3.1.

Dataset	#Conv.	Modality	Labels	Augmentations
MIMIC-CXR-VQA [3]	100,000	X-Ray	VQA	-
Long Tail CXR [14]	88,336	X-Ray	Cls	Seeds,Visual Desc.
Atlas Dataset [15]	82,950	CT	Seg.	Seeds,Visual Prompts
MIMIC-CXR [19]	60,720	X-Ray	Text	Report Filtering
TCGA-UT [20]	30,255	Pathology	Cls	-
Fitzpatrick [10,11]	16,510	Dermatology	Cls	-
Dermnet[a]	15,435	Dermatology	Cls	-
VinDr-CXR [29]	13,123	X-Ray	BB	Groundings
MSD Liver [1]	10,674	CT	Seg.	Spatial Inf.
VinDr-SpineXR [30]	10,340	X-Ray	Cls	Seeds,Visual Desc.
MSD Brain Tumor [1]	10,282	MRI	Seg.	Seeds,Spatial Inf.,Anatomy
MSD HepaticVessel [1]	10,244	CT	Seg.	Seeds,Spatial Inf.,Anatomy
HAM10000 [36]	10,012	Dermatology	Cls	Seeds
SkinCancer [21]	10,002	Pathology	Cls	-
RibFrac [18]	9,221	CT	Seg.	Seeds,Spatial Inf.,Anatomy
VinDr-PCXR [32]	8,939	X-Ray	Cls	Seeds,Visual Desc.
Lung Nodules [2]	8,902	CT	Seg.	Seeds,Spatial Inf.,Anatomy
Pulmonary Embolisms [9]	8,614	CT	Seg.	Seeds,Spatial Inf.,Anatomy
HyperKvasir-Img [5]	8,204	Colonoscopy	Cls	Visual Desc.
GastroVision [17]	7,930	Endoscopy	BB	Groundings
KiTS19 [13]	7,547	CT	Seg.	Seeds,Spatial Inf.,Anatomy
Object-CXR[b]	7,442	X-Ray	BB	Groundings
EyePACS [37]	6,538	Retinal Image	Cls	-
Diabetic Retinopathy[c]	3,540	Retinal Image	Cls	-
VinDr-Mammo [31]	3,396	Mammo	Cls	-
MSD Lung [1]	3,296	CT	Seg.	Seeds,Spatial Inf.,Anatomy
MSD Colon [1]	2,483	CT	Seg.	Seeds,Spatial Inf.,Anatomy
MSD Pancreas [1]	2,117	CT	Seg.	Seeds,Spatial Inf.,Anatomy
HyperKvasir-Seg [5]	1,000	Colonoscopy	BB	Seeds, Groundings
MVID	558,052	Mixed	Vis. Instr.	

[a] https://dermnet.com/
[b] https://github.com/hlk-1135/object-CXR
[c] https://kaggle.com/competitions/diabetic-retinopathy-detection

schedule, and a batch size of 256 for one epoch in Stage 1. In Stage 2, we set the learning rate to $5e^{-6}$. In our second experiment on the full MVID dataset, we incorporate the non-medical LLaVA-1.5 alignment dataset as well as the LLaVA-v1.5 instruction dataset. All experiments were conducted on a single NVIDIA DGX node with 8×H100 GPUs.

4.2 Benchmarks

We evaluate the models on four medical benchmarks: VQA-RAD [23], SLAKE [25], PATH-VQA [12], and GMAI-MMBench [7]. VQA-RAD and SLAKE focus on radiology, while PATH-VQA covers pathology images. These datasets include both closed-ended (yes/no) and open-ended questions. Previous evaluations using word recall [24] tend to favor longer answers, while word-based F1 scores were inconsistent [6]. To enhance reliability, we convert open-ended questions into multiple-choice QA (MCQA) format by using an LLM to generate three plausible distractors. This approach preserves question complexity while ensuring a more precise evaluation. We also include GMAI-MMBench for its broad coverage, spanning 284 datasets across 38 medical imaging modalities, 18 clinical tasks, and 18 departments. It consists entirely of MCQA samples.

We assess model accuracy by prompting VLMs to begin their answers with either the option of a closed-ended question or the corresponding letter of a multiple-choice response. We then compare the model's output to the ground truth to compute the accuracy.

Table 2. The results of our comparison of models trained on LLaVA-Med and MVID60k . We report the accuracy on closed and multiple-choice question answering (MCQA).

Data	Encoder	VQA-RAD Closed	VQA-RAD MCQA	SLAKE Closed	SLAKE MCQA	PATH-VQA Closed	PATH-VQA MCQA	GMAI	AVG
Mistral-7B-Instruct-v0.2									
LLaVA-Med	Clip336	55.38	34.64	62.54	46.20	55.52	36.93	29.00	45.74
MVID60k	Clip336	57.77	32.40	63.70	50.23	59.88	40.09	33.55	48.23
MVID60k	BioClip	61.35	34.70	65.38	51.28	58.66	39.63	37.83	49.83
Biomistral-7B									
MVID60k	Clip336	57.41	46.93	62.74	60.16	57.29	41.36	31.87	51.11
MVID60k	BioClip	59.73	49.16	66.11	62.48	59.19	42.78	36.88	53.76
Qwen2.5-VL-7B-Instruct									
MVID60k	Clip336	64.54	51.96	62.98	69.30	59.58	48.61	37.31	56.33
Llava-Med	BioClip	64.94	55.31	58.41	64.19	55.38	49.31	41.30	55.55
MVID60k	BioClip	68.53	55.87	72.60	69.92	59.73	45.52	42.57	59.25

4.3 Results

Experiment 1. Table 2 presents the results of our first experiment. Training on MVID60k using the same configuration as LLaVA-Med-v.1.5 led to improvements in six out of seven metrics, with an average performance gain of 2.49%. Among image encoders, BioCLIP (224 × 224 resolution) outperformed the general-domain CLIP model (336 × 336 resolution), highlighting the advantages of a specialized visual encoder on biomedical tasks. Despite being based on the weaker Mistral-7B-Instruct-v0.1, the biomedical LLM Biomistral performed better, particularly when paired with BioCLIP, suggesting a synergistic effect between the two. However, the strongest overall model was Qwen2.5-7B-Instruct, a more recent general-domain LLM. Leveraging these insights, we achieved a **13.51% average performance improvement** over the initial LLaVA-Med-v1.5 checkpoint while using the same amount of data. A performance gap of 3.7% remains when training a model with the same configuration on the LLaVA-Med dataset. Additionally, our MCQA reformulations appear to enhance expressiveness, particularly in PATH-VQA, where the MCQA version produced a wider range of results compared to the closed-question format.

Experiment 2. We present the results of our second experiment in Table 3. By training on the entire MVID dataset, we increase the **average performance improvement to LaVA-Med-v1.5 to 20.29%**. The model trained on MVID outperforms most general and medical VLMs included in our comparison and surpasses GPT-4o on VQA-RAD and the closed-question subset of SLAKE. HuatuoGPT-Vision-7B [6] performs better on non-radiological tasks, however, it was trained on a large dataset generated by GPT-4 Vision, an already strong VLM. In contrast, our approach is based on a text-only, open-weight LLM,

Table 3. The results of our comparison with between our best model and different existing VLMs.

Model	VQA-RAD		SLAKE		PATH-VQA		GMAI	AVG
	Closed	MCQA	Closed	MCQA	Closed	MCQA		
General VLMs								
LLaVA-v1.6-7B [26]	52.99	35.75	54.81	48.22	51.61	46.03	35.44	48.23
LLaVA-v1.6-13B [26]	61.35	38.55	54.09	55.04	54.91	47.21	36.63	51.86
LLaVA-v1.6-34B [26]	63.35	55.31	56.49	66.20	64.40	49.24	38.19	59.16
GPT4o	66.53	64.25	71.63	74.57	**75.97**	52.49	**53.96**	67.57
Medical VLMs								
Med-Flamingo [28]	54.18	25.70	52.88	21.86	56.81	23.75	11.64	39.20
Llava-Med v1.5[24]	55.38	34.64	62.54	46.20	55.52	36.93	29.00	45.74
HuatuoGPT-Vision-7B [6]	70.59	64.80	**75.42**	**75.66**	59.29	**54.55**	50.55	67.16
MVID-Qwen2.5-7B	**72.06**	**65.36**	73.25	73.49	61.54	50.47	44.06	66.03

demonstrating competitive performance despite more constrained and privacy preserving resources.

5 Discussion

Our results demonstrate the feasibility of training a competitive medical VLM by leveraging labeled medical imaging data and an open-weight LLM. This finding is particularly relevant in clinical research environments, where data privacy restrictions usually prohibit the use of commercial LLMs. In such settings, researchers frequently have access to labeled imaging datasets, making our approach a viable alternative to proprietary solutions.

Furthermore, our study highlights that despite not utilizing a stronger VLM for data generation, our model's performance remains comparable to those that integrate GPT-based VLMs. This suggests that an open-weight LLM, when combined with labeled imaging data, can effectively support medical VLM training, denying the dependence on proprietary models.

One notable observation is the performance bias towards radiological tasks, which may stem from their overrepresentation in MVID. In contrast, for other imaging modalities, our model often relied on classification label-based information due to the limited availability of more detailed annotations. This suggests that expanding labeled datasets across diverse medical imaging domains could further enhance the model generalizability.

In future work, extending our approach with an open-weight VLM could provide additional benefits. By incorporating a more advanced vision model, we anticipate improvements in both multimodal understanding and task performance, particularly in underrepresented imaging modalities. Additionally, investigating strategies to balance dataset composition and annotation quality across modalities could mitigate performance biases and enhance model robustness in broader clinical applications.

6 Conclusion

In conclusion, our results demonstrate that it is possible to train competitive medical vision-language models using only labeled imaging datasets, without relying on proprietary LLMs. By leveraging MVID, a dataset generated with an open LLM, we show that medical VLMs achieve strong performance on medical VQA benchmarks and perform competitively with state-of-the-art VLMs.

References

1. Antonelli, M., et al.: The medical segmentation decathlon. Nature Commun. **13**(1), 4128 (2022)
2. Armato III, S.G., et al.: Data from LIDC-IDRI a completed reference database of lung nodules on CT scans. Cancer Imaging Arch. (2015)

3. Bae, S., et al.: EHRXQA: a multi-modal question answering dataset for electronic health records with chest x-ray images. In: Thirty-seventh Conference on Neural Information Processing Systems Datasets and Benchmarks Track (2023)
4. Billot, B., et al.: Synthseg: segmentation of brain MRI scans of any contrast and resolution without retraining. Med. Image Anal. **86**, 102789 (2023)
5. Borgli, H., et al.: HyperKvasir, a comprehensive multi-class image and video dataset for gastrointestinal endoscopy. Sci. Data **7**(1), 283 (2020)
6. Chen, J., et al.: Towards injecting medical visual knowledge into multimodal LLMs at scale. In: Proceedings of the 2024 Conference on Empirical Methods in Natural Language Processing, pp. 7346–7370 (2024)
7. Chen, P., et al.: Gmai-mmbench: a comprehensive multimodal evaluation benchmark towards general medical AI. arXiv preprint arXiv:2408.03361 (2024)
8. Dubey, A., et al.: The llama 3 herd of models. arXiv preprint arXiv:2407.21783 (2024)
9. González, G., et al.: Computer aided detection for pulmonary embolism challenge (cad-pe). arXiv preprint arXiv:2003.13440 (2020)
10. Groh, M., et al.: Evaluating deep neural networks trained on clinical images in dermatology with the fitzpatrick 17k dataset. In: Proceedings of the IEEE/CVF Conference on Computer Vision and Pattern Recognition, pp. 1820–1828 (2021)
11. Groh, M., et al.: Towards transparency in dermatology image datasets with skin tone annotations by experts, crowds, and an algorithm. Proc. ACM Human-Comput. Interact. **6**(CSCW2), 1–26 (2022)
12. He, X., et al.: Pathvqa: 30000+ questions for medical visual question answering. arXiv preprint arXiv:2003.10286 (2020)
13. Heller, N., et al.: The kits19 challenge data: 300 kidney tumor cases with clinical context, CT semantic segmentations, and surgical outcomes. arXiv preprint arXiv:1904.00445 (2019)
14. Holste, G., et al.: Long-tailed classification of thorax diseases on chest x-ray: A new benchmark study. In: MICCAI Workshop on Data Augmentation, Labelling, and Imperfections, pp. 22–32. Springer (2022)
15. Jaus, A., et al.: Towards unifying anatomy segmentation: Automated generation of a full-body CT dataset via knowledge aggregation and anatomical guidelines. arXiv preprint arXiv:2307.13375 (2023)
16. Jeong, D.P., et al.: Medical adaptation of large language and vision-language models: are we making progress? arXiv preprint arXiv:2411.04118 (2024)
17. Jha, D., et al.: Gastrovision: a multi-class endoscopy image dataset for computer aided gastrointestinal disease detection. In: ICML Workshop on Machine Learning for Multimodal Healthcare Data (ML4MHD 2023) (2023)
18. Jin, L., et al.: Deep-learning-assisted detection and segmentation of rib fractures from CT scans: Development and validation of fracnet. eBioMedicine (2020)
19. Johnson, A.E., et al.: Mimic-CXR, a de-identified publicly available database of chest radiographs with free-text reports. Sci. data **6**(1), 317 (2019)
20. Komura, D., et al.: Universal encoding of pan-cancer histology by deep texture representations. Cell Rep. **38**, 110424 (2022)
21. Kriegsmann, K., et al.: Deep learning for the detection of anatomical tissue structures and neoplasms of the skin on scanned histopathological tissue sections. Front. Oncol. **12**, 1022967 (2022)
22. Labrak, Y., et al.: Biomistral: a collection of open-source pretrained large language models for medical domains. In: Findings of the Association for Computational Linguistics ACL 2024, pp. 5848–5864 (2024)

23. Lau, J.J., et al.: A dataset of clinically generated visual questions and answers about radiology images. Sci. data **5**(1), 1–10 (2018)
24. Li, C., et al.: LLaVA-med: training a large language-and-vision assistant for biomedicine in one day. In: Thirty-seventh Conference on Neural Information Processing Systems Datasets and Benchmarks Track (2023)
25. Liu, B., et al.: Slake: a semantically-labeled knowledge-enhanced dataset for medical visual question answering. In: 2021 IEEE 18th International Symposium on Biomedical Imaging (ISBI), pp. 1650–1654. IEEE (2021)
26. Liu, H., et al.: Improved baselines with visual instruction tuning. In: Proceedings of the IEEE/CVF Conference on Computer Vision and Pattern Recognition, pp. 26296–26306 (2024)
27. Moor, M., et al.: Foundation models for generalist medical artificial intelligence. Nature **616**(7956), 259–265 (2023)
28. Moor, M., et al.: Med-flamingo: a multimodal medical few-shot learner. arXiv preprint arXiv:2307.15189 (2023)
29. Nguyen, H.Q., et al.: Vindr-CXR: an open dataset of chest x-rays with radiologist's annotations. Sci. Data **9**(1), 429 (2022)
30. Nguyen, H.T., et al.: VinDr-SpineXR: a deep learning framework for spinal lesions detection and classification from radiographs. In: de Bruijne, M., Cattin, P.C., Cotin, S., Padoy, N., Speidel, S., Zheng, Y., Essert, C. (eds.) MICCAI 2021. LNCS, vol. 12905, pp. 291–301. Springer, Cham (2021). https://doi.org/10.1007/978-3-030-87240-3_28
31. Nguyen, H.T., et al.: Vindr-mammo: a large-scale benchmark dataset for computer-aided diagnosis in full-field digital mammography. Sci. Data **10**(1), 277 (2023)
32. Pham, H.H., et al.: Vindr-PCXR: an open, large-scale pediatric chest x-ray dataset for interpretation of common thoracic diseases. PhysioNet (2022)
33. Radford, A., et al.: Learning transferable visual models from natural language supervision. In: International Conference on Machine Learning, pp. 8748–8763. PMLR (2021)
34. Saab, K., et al.: Capabilities of Gemini models in medicine. arXiv preprint arXiv:2404.18416 (2024)
35. Sun, G., et al.: Self-training large language and vision assistant for medical question answering. In: Proceedings of the 2024 Conference on Empirical Methods in Natural Language Processing, pp. 20052–20060 (2024)
36. Tschandl, P., et al.: The ham10000 dataset, a large collection of multi-source Dermatoscopic images of common pigmented skin lesions. Sci. Data **5**(1), 180161 (2018)
37. de Vente, C., et al.: Airogs: artificial intelligence for robust glaucoma screening challenge. arXiv preprint arXiv:2302.01738 (2023)
38. Wasserthal, J., et al.: Totalsegmentator: robust segmentation of 104 anatomic structures in CT images. Radiol. Artif. Intell. **5**(5), e230024 (2023)
39. Yang, A., et al.: Qwen2. 5 technical report. arXiv preprint arXiv:2412.15115 (2024)
40. Zhang, S., et al.: A multimodal biomedical foundation model trained from fifteen million image–text pairs. NEJM AI **2**(1) (2024)

DR.SIMON: Domain-Wise Rewrite for Segment-Informed Medical Oversight Network

Seohyun Lee[1](✉), Suhyun Choe[2], Jaeha Choi[3], and Jin Won Lee[4]

[1] Korea University, Seoul, Republic of Korea
happy8825@korea.ac.kr
[2] Yonsei University, Seoul, Republic of Korea
gydbs0925@yonsei.ac.kr
[3] Incheon National University, Incheon, Republic of Korea
chlgocks2000@inu.ac.kr
[4] McGill University, Montreal, Canada
jinwon.lee@mail.mcgill.ca

Abstract. Humans are capable of understanding language, even when encountering unfamiliar words. Rather than requiring precise definitions, we often infer meaning from the surrounding linguistic context or visual cues. Inspired by this capability, we address the long-standing challenge of aligning medical terminology in queries with visual content for temporal grounding in medical videos. While bridging this gap typically relies on costly, domain-specific fine-tuning, such methods frequently lack generalization and struggle to adapt to newly coined or rarely encountered terms. To deal with this limitation, we present **DR.SIMON** (**D**omain-wise **R**ewrite for **S**egment-**I**nformed **M**edical **O**versight **N**etwork), a simple yet efficient query-rewriting framework that runs on a frozen backbone. DR.SIMON first segments the video into coarse events, then rewrites the user query into visually explicit paraphrases under global visual context, and finally localizes the most relevant segment. Evaluated on MedVidCL, DR.SIMON achieves remarkable gains over recent video-LLMs—without any additional training. Our results show that mitigating lexical misalignment alone can unlock substantial performance improvements and provide a scalable route to keep pace with continually emerging medical vocabulary. Code will be released at https://drsimon-rewrite.github.io/.

Keywords: Video temporal grounding · Domain specific query rewriting · Video event segmentation

1 Introduction

The rapid advancement of vision-language model (VLM) has significantly accelerated progress in video understanding tasks [13,15,17,22]. In particular, video

---

S. Choe, J. Choi and J. W. Lee—Contributed equally as second authors, ordered alphabetically.

(a) Original query

Query : During which frames can we see how to treat **patellofemoral pain** by performing straight side leg raises?

(b) Masked query

Query : During which frames can we see how to treat **something** by performing straight side leg raises?

Fig. 1. Diagnostic Masking Procedure. (a) Original query containing a domain-specific medical term (e.g., *polypectomy*). (b) The same query after replacing that term with the neutral placeholder token "something". This isolates the influence of the medical term and forms the basis of the ablation in Table 1.

temporal grounding (VTG) localizes segments in a video and underpins downstream tasks such as retrieval and summarization [6,9,12]. While VTG has been widely investigated in general-domain videos [4,6,8,11,19,25], it remains challenging in domain-specific videos, particularly medical videos, due to their long durations and the need for expert-level semantic and terminological understanding. Recent approaches address this by integrating external medical knowledge graphs or structured report data [5,18,24]. However, these methods typically require extensive training and large-scale domain-specific datasets.

To determine how much the lexical gap hurts grounding, we perform a simple ablation (Fig. 1) in which every domain-specific medical term in queries is masked. Counter-intuitively, overall performance improves (Table 1). The improvement is small yet revealing: the limiting factor is not visual perception but the mismatch between domain-specific terminology and the generic visual cues on which the model was trained. This motivates our key idea: rewrite the query into visually explicit form before grounding.

In light of this, we propose DR.SIMON, a simple yet efficient framework that rewrite the query itself into visually explicit language, achieving comparable gains with significantly lower cost. Without additional training, DR.SIMON improves VTG in long-form medical videos in three stage process. (1) Query rewriting. It employs a frozen Video-LLM to generate a visually explicit paraphrase of the medical question, augmenting the medical query with observable actions. (2) Segmentation. It divides the input video into semantically coherent clips and extracts representative actions of each segment. (3) Selection and localization. It computes similarity between each rewritten query and actions extracted from the video and selects the most relevant segment.

Our experimental results demonstrate that DR.SIMON outperforms existing methods without any training overhead, showing that query rewriting alone effectively mitigates linguistic–visual misalignment in medical video understanding.

2 Related Works

2.1 Temporal Grounding in Video-Language Models

Temporal grounding locates the start and end times of an event in a video given a natural-language query. Early studies focused on generating high-level summaries of entire videos but struggled to perform fine-grained temporal grounding. Zhang et al. [26] addresses this challenge by introducing a model, Temporal Query Networks, which allows fine-grained action recognition and temporal localization in untrimmed videos.

With the advent of large language models (LLMs), recent studies have significantly improved grounding performance. VTimeLLM [11] employs a boundary-aware training pipeline for precise event localization; however, it relies on fixed-frame sampling and short context windows, making it challenging to handle long videos. ReVisionLLM [8] addresses this limitation by introducing a hierarchical, recursive strategy, requiring extensive hierarchical training on large-scale video datasets. While effective for general-domain videos, this approach necessitates additional domain-specific fine-tuning to accurately interpret specialized terminology, such as medical queries. To address these limitations we employ a sliding-window segmentation approach to efficiently handle long videos. Also we reformulate temporal grounding as an event-matching problem, significantly reducing computational overhead and effectively handling domain-specific terminology without additional training.

2.2 Query Reformulation and Handling Domain-Specific Terminology

Rewriting queries (e.g., acronym expansion, simplification) improves alignment between language and video modalities. Sun et al. [23] employ an LLM (MiniGPT-4 [27]) to rephrase a user's query into an enriched version and to generate detailed textual descriptions of video frames. The system improves moment retrieval performance by matching the rewritten queries with frame-level descriptions. However, this text-only rewriting method is blind to the actual video content and, without domain-specific fine-tuning, often fails to handle specialized medical terms. In the medical QA context, handling domain-specific terminology is a key challenge. Recent approaches explore ways to make medical queries more interpretable to models. For example, Cho and Lee [3] propose a method that detects medical terms in a question and injects definitions or explanations of those terms, thereby enabling a general LLM to better grasp the meaning of the question. Although effective, these strategies overlook visual context which limits cross-domain scalability and overall performance. By contrast, our framework explicitly rewrites the query under global visual context. The Video-LLM jointly attends to the video and the original query, injects visually observed actions into the sentence, and produces an action-centric query rewrite. This video-guided rewriting eliminates reliance on external knowledge bases, bridges the lexical gap even for unseen terms, and scales to diverse domains without additional training.

3 Method

Overview. We perform temporal grounding of a natural language query in an untrimmed video as illustrated in Fig. 2. Section 3.1 rewrites the original query q_0

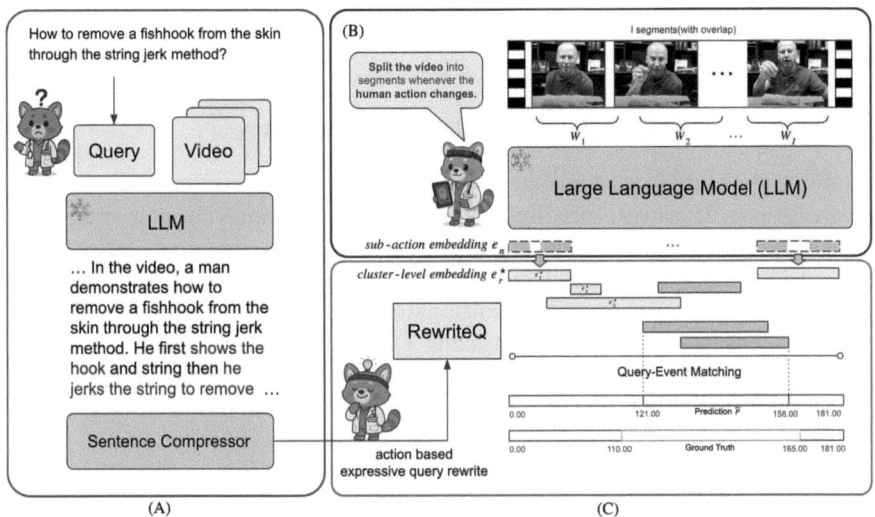

Fig. 2. Overall pipeline of DR.SIMON. (A) Rewrite the query into visually explicit language. (B) Slice the video into overlapping windows and extract representative events. (C) Match the rewritten query to those events and output the most relevant time span $\widehat{P}$. Visual encoder and VtimeLLM adapter is omitted for brevity.

using global visual context, yielding an action-centric query $\hat{q}$ that better aligns with the video's semantics. Section 3.2 sweeps an overlapping window across the video, clusters the resulting action proposals, and refines their start–end boundaries to yield a concise set of representative events' embedding $e_i^\star$. Section 3.3 ranks the representative events by cosine similarity and selects the most coherent high-score cluster; the outermost boundaries of this cluster constitute the predicted interval $\widehat{P}$.

3.1 Query Rewriting Module (QRM)

Given an untrimmed video V of duration T seconds, we extract a sequence of frame-level CLIP [20] embeddings, denoted by $\mathbf{F} = [\mathbf{f}_1, \ldots, \mathbf{f}_N] \in \mathbb{R}^{N \times d}$, where N is the total number of frames. In parallel, the original query q_0 is processed by the CLIP [20] text encoder to obtain the initial textual embedding $\mathbf{q}_0$. To obtain an action-centric rewritten query, we first pass the frame embeddings $\mathbf{F}$ through the frozen VTimeLLM [11] visual adapter to convert them into LLM-readable visual tokens. These tokens are then given as input to a frozen LLM, which produces the action-focused description $\tilde{q}$. Since $\tilde{q}$ can contain repetitive or semantically redundant phrases, we decompose it into individual sentences. We embed each with a Sentence-BERT [21] encoder and merge those whose pairwise cosine similarity exceeds a predefined threshold τ. This processing produces a concise reformulation $\hat{q}$, a clear action-driven query, that better aligns with the video. Notably, enabling QRM by itself already boosts performance (see "Rewrite Query (Summ.)" in Table 1).

3.2 Boundary Event Segmentation Module (BESM)

To effectively handle action-centric queries, we require precise, action-oriented video captions. However, directly generating these captions faces practical challenges due to overly fragmented outputs. For instance, feeding entire videos into a Video-LLM often produces excessively detailed micro-events accompanied by repetitive captions [14]. Additionally, large language models suffer from positional biases, notably under-utilizing information located in the middle of lengthy inputs [16]. Consequently, we adopt a sliding-window strategy followed by density-based clustering, which robustly merges redundant sub-actions and yields one representative caption per semantic event.

The video is first divided into I overlapping windows $W_1, \ldots, W_I$ using a fixed stride and overlap ratio. For every window W_i, VLM model returns a set of sub-action captions $\mathcal{A}_i = \{a_{ij}\}_{j=1}^{m_i}$, where $m_i = |\mathcal{A}_i|$, each with start/end percentages that are converted into absolute seconds inside W_i. Encoding every caption once with Sentence-BERT yields the unit-norm vectors $\mathcal{E}_i = \{\mathbf{e}_{ij}\}_{j=1}^{m_i} \subset \mathbb{R}^d$.

All sub-actions from all windows are pooled into $\{(t_n^{\text{start}}, t_n^{\text{end}}, \mathbf{e}_n)\}_{n=1}^{M}$, where $M = \sum_i m_i$ and $t_n^{\text{start}}, t_n^{\text{end}}$ are absolute start/end times. We construct a pairwise cosine-distance matrix $D_{nm} = 1 - \mathbf{e}_n^\top \mathbf{e}_m$ and run average-linkage agglomerative clustering with threshold $1 - \tau$. The influence of the density threshold τ is analyzed in Sect. 4.3.

Each resulting cluster C_r yields the earliest start, $\min_{n \in C_r} t_n^{\text{start}}$, and the latest end $\max_{n \in C_r} t_n^{\text{end}}$. The caption whose embedding is closest to the cluster centroid[1]. We denote this representative caption by $a_r^\star$ and its embedding by $\mathbf{e}_r^\star$. The sequence $\{(a_r^\star, \mathbf{e}_r^\star)\}_{r=1}^{R}$, where $R = |C_r|$, which contains one event for each semantic cluster, is forwarded to Sect. 3.3 for query–event matching.

3.3 Query–Event Matching Module (QEM)

We reformulate the VTG task as an event-matching problem where an action-centric query is aligned with event-level captions. This allows the model to focus on video segments most likely to contain the answer. Given the rewritten query $\hat{q}$, we embed it as a vector $\mathbf{q}$ and compute cosine similarities $\text{score}_i = \mathbf{q}^\top \mathbf{e}_r^\star$ for every event embedding $\mathbf{e}_r^\star$. Instead of scanning the full video, we retain the seven highest-scoring events, ordered on the timeline $t_{a_1^\star}^{\text{start}} \prec \ldots \prec t_{a_7^\star}^{\text{start}}$. Focusing on top-7 strip drastically reduces the search space.

When the three best events already form a compact chunk—each pair either overlaps (IoU > 0.2) or their temporal gap is at most $\theta_{\text{gap}} = 1.5\,\text{s}$—we treat them as a single action and return the union of their boundaries. This covers the majority of queries in which the relevant frames are naturally adjacent.

---

[1] $\hat{\mathbf{e}}_r = \frac{1}{|C_r|} \sum_{n \in C_r} \mathbf{e}_n, \quad \mathbf{e}_r^\star = \arg\max_{\mathbf{e}_n : n \in C_r} \hat{\mathbf{e}}_r^\top \mathbf{e}_n.$

If the top–3 events are scattered, a stronger discriminator is needed. We merge adjacent segments whose gaps do not exceed θ_{gap}, yielding clusters $\{C_r\}_{r=1}^R$. For each cluster, we define

$$S_r = (1 + \lambda \rho_r) \frac{1}{|C_r|} \sum_{g \in C_r} \text{score}_g, \qquad \rho_r = \max\left(0, 1 - \frac{\text{span}(C_r)}{|C_r|\theta_{\text{gap}}}\right), \qquad (1)$$

with $\lambda = 0.1$ and $\text{span}(C_r) = \max_{g \in C_r} t_g^{\text{end}} - \min_{g \in C_r} t_g^{\text{start}}$. The term $\rho_r \in [0, 1]$ rewards tightly packed clusters: it approaches 1 when segments abut and falls to 0 for diffuse groups, thus modulating the average similarity by the density of the cluster. The cluster with the highest score, $r^\star = \arg\max_r S_r$, defines the prediction

$$\widehat{P} = \left[\min_{g \in C_{r^\star}} t_g^{\text{start}}, \max_{g \in C_{r^\star}} t_g^{\text{end}} \right]. \qquad (2)$$

4 Experiments

Experiments are carried out on the MedVidCL [7] test split, which contains 50 medical instructional clips. The average video length is 345.4 s, ranging from 46 s to 20 min. Across these clips there are QA pairs, which requires the model to precisely localize answer segments within long instructional videos. We follow the conventional metrics in VTG: mean Intersection-over-Union (mIoU) and recall at IoU thresholds 0.3, 0.5, and 0.7. A prediction is correct if its IoU with the ground-truth span exceeds the threshold.

4.1 Experiment Configuration

All methods except VSL-QGH [7] share a Vicuna-7B [2] backbone with a CLIP-ViT/L-14 [20], and process videos uniformly subsampled at 2.0fps. We compare seven methods: (1) VTimeLLM [11] (2) VTimeLLM with Masked Query, obtained by replacing every medical term in the query with ``something''; (3) VTimeLLM with Rewrite Query, which inserts a query-rewriting module; (4) VTimeLLM with Summarized Query, which further applies an abstractive summary to the rewritten query; (5) VSL-QGH, a span-prediction baseline enhanced with query-guided highlighting and fine-tuned on the MedVidQA train split; (6) ReVisionLLM [8] and (7) DR.SIMON (ours).

4.2 Results and Analysis

Table 1 presents the quantitative results, highlighting three key observations. (1) Domain specific term harms performance, if the model isn't fine-tuned to the domain. Comparing Row 1 with Row 2, masking medical terms lifts all metrics over the backbone, showing that VTimeLLM [11] rarely grounds such vocabulary visually. (2) Query rewriting helps coarse localization. Feeding a visually explicit rewrite (Row 3) boosts R@0.3, confirming that linguistic alignment is the primary bottleneck. Compressing the rewrite query (Row 4) further

Table 1. Temporal localization results on the MedVidQA test set. Best score in each column is shown in **bold**, and the second-best is underlined.

Method	mIoU ↑	R@0.3 ↑	R@0.5 ↑	R@0.7 ↑
VTimeLLM	6.32	9.86	4.93	2.11
+ Masked Query	7.92	11.97	6.34	3.52
+ Rewrite Query	7.15	10.56	5.63	3.52
+ Rewrite Query (Summ.)	9.18	15.49	7.04	1.41
VSL-QGH	20.12	25.81	14.20	6.45
RevisionLLM	21.18	28.50	**26.10**	**14.28**
DR.SIMON (ours)	**28.08**	**40.14**	20.42	10.56

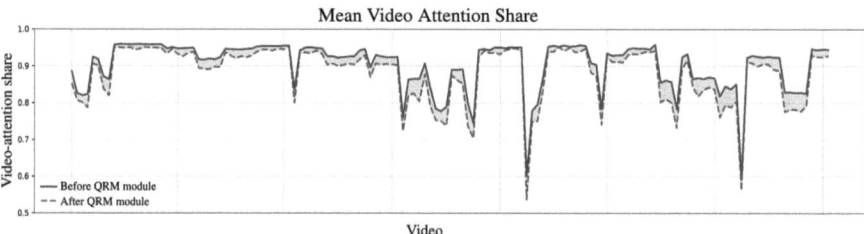

Fig. 3. Cross-modal attention ratio before and after query rewriting. The y-axis is normalized so that 1.0 corresponds to attention devoted exclusively to video tokens, whereas 0.5 indicates a perfectly balanced split between video and text. After QEM module, attention shifts toward the language modality, alleviating the original video bias.

aids coarse windows but erodes strict IoU, suggesting a loss of fine cues. As further evidence, we plot the cross-modal attention ratio in Fig. 3. Before the QEM module is applied, the model exhibits a clear bias toward video information. However, after query rewriting, attention shifts more toward the language modality, producing a more balanced output and thus reducing modal bias. By rewriting the query to match latent video features, the model can rely on textual cues that point precisely to the relevant segment. Figure 3 mirrors the findings of Chen et al. [1]: rephrasing the question while preserving its meaning weakens single-modality shortcuts. (3) DR.SIMON (Row 7) produces the highest mIoU and R@0.3 without any fine-tuning. By enforcing explicit action boundaries, rewriting queries with visual context, and ranking candidate windows by density-weighted relevance, DR.SIMON achieves +6.9%p mIoU and +11.6%p R@0.3 over the strongest baseline in one pass.

4.3 Hyper-Parameter Sensitivity

To quantify hyper-parameter sensitivity in QEM and BESM, we analyze two key parameters: the cluster-density threshold τ, which determines how similar

Table 2. Impact of cluster density τ and fallback pool k. τ is the similarity threshold for merging segments, and k is the number of extra top-scoring segments checked when the initial top-3 are not contiguous.

Varying τ at $k=7$					Varying k at $\tau \in \{0.99, 0.90\}$				
τ	mIoU↑	R@0.3↑	R@0.5↑	R@0.7↑	τ	k	mIoU↑	R@0.3↑	R@0.5↑	R@0.7↑
1.0	22.41	32.39	12.68	7.04	0.99	10	25.99	30.28	15.49	7.75
0.99	**28.08**	**40.14**	**20.42**	**10.56**		7	**28.08**	**40.14**	**20.42**	**10.56**
0.90	25.31	30.99	15.49	9.15		5	23.20	30.28	16.20	10.56
0.80	22.73	24.65	10.56	4.93		3	17.42	23.24	12.00	5.63
0.70	21.53	21.13	9.15	4.23		1	7.65	8.45	3.52	0.70
0.60	19.65	16.20	9.86	4.93	0.90	10	23.94	25.35	13.38	7.75
0.50	18.17	14.08	7.75	3.52		7	25.31	30.99	15.49	9.15
						5	24.86	**33.80**	**17.61**	**9.86**
						3	20.51	26.06	13.38	7.75
						1	8.66	9.86	4.23	0.70

segments must be merged, and the fallback strip size k, the number of additional high-scoring segments reviewed when the initial top-3 test fails to form a compact cluster. Since the MedVidCL [7] dataset contains mostly static, visually similar shots, tightening τ to around 0.99 maintains clusters that are pure yet comprehensive, optimizing both mIoU and recall. Lower thresholds merge less related segments, reducing performance, while excessively high thresholds fragment clusters, diminishing recall (Fig. 2). Moderate fallback sizes (around $k = 7$) notably enhance recall without sacrificing overall precision, whereas larger fallback sizes produce marginal improvements but degrade cluster cohesion.

5 Ablation Studies

5.1 Effect of QRM

To measure how QRM alone aids a backbone not yet domain-tuned, and how that gain shifts after a single-epoch domain fine-tune, we disable BESM and QEM and test both original and rewritten queries. We fine-tune the frozen VTimeLLM [11] with a single-epoch LoRA [10] training on the MedVidQA train split[2]—once using the original queries Base FT and once using our rewritten queries as a form of data augmentation RewriteQ FT.

As shown in Table 3, rewriting remains useful even at train-time. Augmenting the training set with rewritten queries lifts test-time recall by +3.5%p (R@0.3) compared with Base FT, regardless of whether the test query is rewritten. Moreover, we can observe that rewriting at test-time is less critical once the model is domain-tuned. Within each checkpoint the gap between Orig. Q and Rewrite Q is only 0.2–0.3%p mIoU, suggesting the model now understands medical terms. Hence, query rewriting yields its largest benefit when the backbone has not yet been adapted to the domain specific vocabulary.

---

[2] LoRA settings: rank $r=64$, $\alpha=128$; applied to all FFN, self-attention, and cross-attention blocks; batch size 24; 4 × A100-40 GB.

Table 3. Verification of QRM. LoRA fine-tuning results on MEDVIDQA.

Setting	mIoU ↑	R@0.3 ↑	R@0.5 ↑	R@0.7 ↑
Base FT + Orig. Q (Row 1)	14.58	21.13	9.86	4.23
Base FT + Rewrite Q (Row 2)	14.79	21.13	10.56	2.82
RewriteQ FT + Orig. Q (Row 3)	14.99	24.65	9.86	3.52
RewriteQ FT + Rewrite Q (Row 4)	**15.18**	**24.94**	**10.15**	**4.52**

6 Conclusion

We propose **DR.SIMON**, a simple yet efficient framework that achieves superior performance on the MedVidQA dataset without any additional training. We expect rapid application to medical domains with emerging terminology or scarce annotated data, highlighting the power of query reformulation for robust video understanding on medical domains. It also effectively narrows the search space and significantly reduces computational resources by redefining the video temporal grounding task as an event-selection problem. Finally, by injecting visually grounded actions into the rewritten query, DR.SIMON bridges the lexical gap between medical terms and generic video features, reducing the need for domain specific finetuning.

7 Limitations and Future Work

Although our method efficiently achieves remarkable improvements in mIoU and R@0.3 without additional training, it also has certain limitations. Our method relies heavily on the initial segmentation produced by the Video-LLM. If this segmentation results in overly long or overly fragmented segments, or if the boundaries are incorrectly placed, our QEM module may select inaccurate spans. This issue is evident from the performance drop observed at higher IoU thresholds (R@0.5 and R@0.7), despite strong results at lower thresholds (R@0.3). Another limitation is the sensitivity to hyperparameter choices, including window size, overlap ratio, and clustering thresholds. Optimal settings for these parameters vary significantly based on video characteristics.

To address these limitations, we plan to explore finer-grained approaches that can precisely refine segment boundaries after coarse event selection. Additionally, adaptive strategies for automatically tuning hyperparameters according to video content will be investigated to enhance robustness across diverse videos. Lastly, we aim to extend our framework beyond medical videos, establishing a generalizable solution to handle linguistic-visual alignment challenges in other specialized domains.

References

1. Chen, M., Cao, Y., Zhang, Y., Lu, C.: Quantifying and mitigating unimodal biases in multimodal large language models: a causal perspective (2024)
2. Chiang, W.-L., et al.: Vicuna: an open-source chatbot impressing GPT-4 with 90%* ChatGPT quality (2023)
3. Cho, J., Lee, G.: K-comp: retrieval-augmented medical domain question answering with knowledge-injected compressor. In: Proceedings of the 2025 Conference of the Nations of the Americas Chapter of the Association for Computational Linguistics: Human Language Technologies (Volume 1: Long Papers), pages 6878–6901 (2025)
4. Gao, J., Sun, C., Yang, Z., Nevatia, R.: TALL: temporal activity localization via language query. In: 2017 IEEE International Conference on Computer Vision (ICCV), pp. 5277–5285 (2017)
5. Gu, T., Yang, K., Liu, D., Cai, W.: LaPA: latent prompt assist model for medical visual question answering. In: 2024 IEEE/CVF Conference on Computer Vision and Pattern Recognition Workshops (CVPRW), pp. 4971–4980 (2024)
6. Guo, Y., Liu, J., Li, M., Liu, Q., Chen, X., Tang, X.: TRACE: temporal grounding video LLM via causal event modeling (2025)
7. Gupta, D., Attal, K., Demner-Fushman, D.: A dataset for medical instructional video classification and question answering (2022). abs/2201.12888
8. Hannan, T., Islam, M., Gu, J., Seidl, T., Bertasius, G.: ReVisionLLM: recursive vision-language model for temporal grounding in hour-long videos (2024)
9. He, B., Wang, J., Qiu, J., Bui, T., Shrivastava, A., Wang, Z.: Align and Attend: multimodal summarization with dual contrastive losses. In: Proceedings of the IEEE/CVF Conference on Computer Vision and Pattern Recognition (CVPR) (2023)
10. Hu, E., et al.: LoRA: low-rank adaptation of large language models (2021)
11. Huang, B., Wang, X., Chen, H., Song, Z., Zhu, W.: VtimeLLM: empower LLM to grasp video moments. In: 2024 IEEE/CVF Conference on Computer Vision and Pattern Recognition (CVPR), pp 14271–14280 (2024)
12. Islam, M., Hasan, M., Athrey, K., Braskich, T., Bertasius, G.: Efficient movie scene detection using state-space transformers. In: Proceedings of the IEEE/CVF Conference on Computer Vision and Pattern Recognition, pp. 18749–18758 (2023)
13. Jin, P., Takanobu, R., Zhang, W., Cao, X., Yuan, L.: Chat-UniVi: unified visual representation empowers large language models with image and video understanding. In: 2024 IEEE/CVF Conference on Computer Vision and Pattern Recognition (CVPR), pp. 13700–13710 (2024)
14. Lee, J., Kim, J., Na, J., Park, J., Kim, H.: VidChain: chain-of-tasks with metric-based direct preference optimization for dense video captioning. Proc. AAAI Conf. Artif. Intell. **39**, 4499–4507 (2025)
15. Lin, B., et al.: Video-LLaVA: learning united visual representation by alignment before projection (2024)
16. Liu, N., et al.: Lost in the middle: how language models use long contexts. Trans. Assoc. Comput. Linguist. **12**, 157–173 (2024)
17. Liu, S., Zhang, C., Zhao, C., Ghanem, B.: End-to-end temporal action detection with 1b parameters across 1000 frames. In: 2024 IEEE/CVF Conference on Computer Vision and Pattern Recognition (CVPR), pp. 18591–18601 (2024)
18. Nath, V., et al.: VILA-M3: enhancing vision-language models with medical expert knowledge (2025)

19. Qian, L.: Momentor: advancing video large language model with fine-grained temporal reasoning. In: Proceedings of the 41st International Conference on Machine Learning (2024)
20. Radford, A., et al.: Learning transferable visual models from natural language supervision (2021)
21. Reimers, N., Gurevych, I.: Sentence-BERT: sentence embeddings using Siamese BERT-networks. In: Proceedings of the 2019 Conference on Empirical Methods in Natural Language Processing and the 9th International Joint Conference on Natural Language Processing (EMNLP-IJCNLP), pp. 3982–3992 (2019)
22. Ren, S., Yao, L., Li, S., Sun, X., Hou, L.: TimeChat: a time-sensitive multimodal large language model for long video understanding. In: 2024 IEEE/CVF Conference on Computer Vision and Pattern Recognition (CVPR), pp. 14313–14323 (2024)
23. Sun, Y., Xu, Y., Xie, Z., Shu, Y., Du, S.: GPTSee: enhancing moment retrieval and highlight detection via description-based similarity features. IEEE Signal Process. Lett. **31**, 521–525 (2024)
24. Wu, C., Zhang, X., Zhang, Y., Wang, Y., Xie, W.: MedKLIP: medical knowledge enhanced language-image pre-training for X-Ray diagnosis. In: 2023 IEEE/CVF International Conference on Computer Vision (ICCV), pp. 21315–21326 (2023)
25. Wu, Y., et al.: Number it: temporal grounding videos like flipping manga. In: Proceedings of the Computer Vision and Pattern Recognition Conference (CVPR), pp. 13754–13765 (2025)
26. Zhang, C., Gupta, A., Zisserman, A.: Temporal query networks for fine-grained video understanding. In: Proceedings of the IEEE/CVF Conference on Computer Vision and Pattern Recognition (CVPR) (2021)
27. Zhu, D., Chen, J., Shen, X., Li, X., Elhoseiny, M.: MiniGPT-4: enhancing vision-language understanding with advanced large language models. arXiv preprint arXiv:2304.10592 (2023)

The Data Behind the Model: Gaps and Opportunities for Foundation Models in Brain Imaging

Salah Ghamizi[1](✉)[iD], Georgia Kanli[1,2][iD], Magali Perquin[1][iD], and Olivier Keunen[1][iD]

[1] Luxembourg Institute of Health (LIH), Esch-sur-Alzette, Luxembourg
{salah.ghamizi,georgia.kanli,magali.perquin,olivier.keunen}@lih.lu
[2] Faculty of Electrical Engineering and Communication, Brno University of Technology, Brno, Czech Republic

Abstract. Foundation Models (FMs) have revolutionized machine learning in medical imaging, yet their application to brain imaging remains limited and fragmented. Despite the availability of diverse and extensive neuroimaging datasets, most FM research has focused narrowly on a handful of tasks, mainly tumor classification and segmentation, while neglecting prevalent neurological disorders such as ADHD and early-stage Parkinson's disease. In this work, we present the largest and most comprehensive atlas of brain imaging datasets to date, comprising 151 datasets and over 541k volumetric imaging studies across a wide range of modalities and pathologies. Our meta-analysis of 86 brain imaging FMs reveals a disproportionate reliance on structural MRI and a small set of popular datasets, along with critical blind spots in both disease coverage and imaging modalities. We identify systemic challenges, including inconsistent model evaluation protocols, heterogeneous data formats, and limited availability. All of which hinder reproducibility, scalability, and clinical translation. Our publicly available atlases pave the way for more robust, scalable, and clinically meaningful FMs in brain imaging.

Keywords: Foundation Models · Medical Imaging · Brain diseases

1 Introduction

Recent advances in machine learning (ML), particularly the development of Foundation Models (FMs), have marked the beginning of a new era in medical imaging. Foundation models are large-scale models trained on large amounts of data, serving as a foundation for various downstream tasks (CLIP, SAM).

Although FMs have recently tackled medical applications, we argue that brain pathologies have been largely ignored in FM research. For example, [8], one of the largest reviews included 200 technical articles, but only 32 covered brain pathologies; Similarly, the study by Shi et al. [18] provided one of the most

---

S. Ghamizi and G. Kanli—These authors contributed equally.

© The Author(s), under exclusive license to Springer Nature Switzerland AG 2026
W.-K. Jeong et al. (Eds.): MedAGI 2025, LNCS 16112, pp. 109–119, 2026.
https://doi.org/10.1007/978-3-032-07845-2_11

comprehensive analyzes and covered 76 research articles, but only ten dealt with brain imaging.

Furthermore, even FMs tackling brain imaging only focus on "popular" ML tasks such as tumor classification or segmentation, missing prevalent neurological disorders like ADHD and early Parkinson predictions. Contrary to the common belief that brain imaging research lacks diverse high-quality datasets and benchmarks to adequately design and evaluate brain FM, we argue on the contrary that (1) a rich variety of brain imaging datasets is already available, (2) current model design and evaluation protocols are not consistent, making fair comparison between FMs difficult, and (3) there is a disconnect between translational applications of medical machine and the main research direction for brain FM.

We support our claims with an evaluation of 151 brain imaging datasets and 86 brain imaging FMs. Our contributions can be summarized as follows.

- We offer the largest brain dataset atlas, compiling imaging datasets totaling more than 541k volumetric studies.
- We provide the largest curation of brain FM, and offer a dataset-oriented meta-analysis.
- We discuss and highlight the main trends and common pitfalls in FM research for brain imaging, and offer a set of recommendations and good practices to start brain imaging FM research.

Given the restrictions on the number of reference pages, we provide complete references of the datasets and models in the atlas.

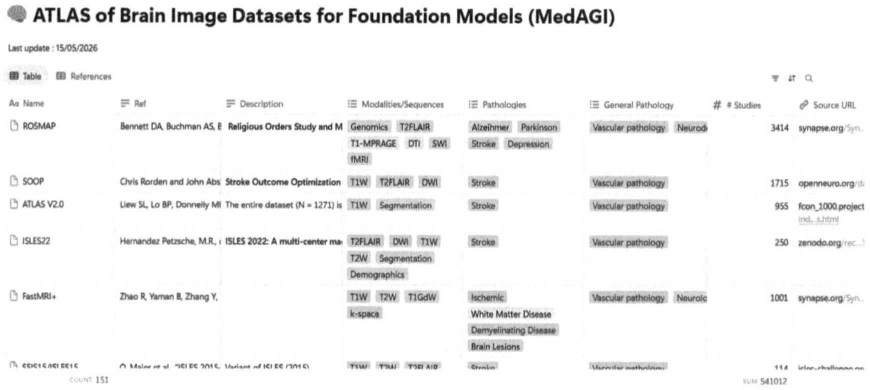
ATLAS of Brain Image Datasets for Foundation Models (MedAGI)

Last update : 15/05/2026

Table References

Name	Ref	Description	Modalities/Sequences	Pathologies	General Pathology	# Studies	Source URL
ROSMAP	Bennett DA, Buchman AS, B	Religious Orders Study and M	Genomics T2FLAIR T1-MPRAGE DTI SWI fMRI	Alzeihmer Parkinson Stroke Depression	Vascular pathology Neurod	3414	synapse.org/Syn..
SOOP	Chris Rorden and John Abs	Stroke Outcome Optimization	T1W T2FLAIR DWI	Stroke	Vascular pathology	1715	openneuro.org/d
ATLAS V2.0	Liew SL, Lo BP, Donnelly MI	The entire dataset (N = 1271) is	T1W Segmentation	Stroke	Vascular pathology	955	fcon_1000.project ind...s.html
ISLES22	Hernandez Petzsche, M.R.,	ISLES 2022: A multi-center ma	T2FLAIR DWI T1W T2W Segmentation Demographics	Stroke	Vascular pathology	250	zenodo.org/rec...1
FastMRI+	Zhao R, Yaman B, Zhang Y,		T1W T2W T1GdW k-space	Ischemic White Matter Disease Demyelinating Disease Brain Lesions	Vascular pathology Neurolo	1001	synapse.org/Syn..
SSIS15/ISLES15	O. Maier et al. "ISLES 2015.	Variant of ISLES (2015)	T1W T2W T2FLAIR	Stroke	Vascular pathology	114	isles-challenge.or
COUNT 151						SUM 541012	

Fig. 1. Our interactive atlas is publicly available, and actively updated: https://ballistic-purple-6f5.notion.site/176bf710ca5980d2bccdd6029523f3ed

2 Related Work

Foundation Models (FM) are large-scale machine learning models trained on diverse datasets, often unlabeled, using self-supervised learning techniques. They are designed to learn general representations that can be adapted to a wide range of downstream tasks through fine-tuning. Unlike traditional models built for specific applications, FMs offer scalability, transferability, and efficiency, making them powerful tools for complex domains such as medical imaging.

FMs have gained momentum in medical imaging because of their ability to generalize across tasks with minimal fine-tuning. Pioneering models such as MedCLIP, BioViL, or MONAI's generative tools have adapted vision language and vision-only pre-training strategies to handle multimodal clinical data. Although these models demonstrate strong performance in tasks such as generating chest radiographs [21], tumor segmentation [17], or improving medical image quality [2,9] most have limited applicability to 3D imaging modalities such as MRI or CT, and few explore domain-specific tasks tailored to neuroimaging.

Only a small fraction of medical FMs explicitly target brain imaging. Existing FM focus primarily on brain tumor segmentation or Alzheimer's diagnosis. These tasks, while important, represent a limited view of brain-related disorders. Neurological and psychiatric conditions such as ADHD, multiple sclerosis, and Parkinson's disease are often overlooked. In addition, most studies use limited datasets or lack external validation, reducing generalizability.

To the best of our knowledge, there are no studies on brain datasets for FMs. Curating and analyzing large medical datasets relevant to designing and improving FM for brain research is a critical for FM research, and we believe that our work bridges the gap between datasets publishers, and machine learning researchers, leading the way to more generalizable and reliable brain FM.

3 Methodology

This work aims to provide a comprehensive, structured, and accessible atlas of available brain imaging datasets to support the development and evaluation of brain-specific foundation models. To achieve this, we designed a systematic pipeline to identify, curate, classify, and analyze existing datasets relevant to brain imaging tasks.

3.1 Collection, Identification and Inclusion Criteria

We began by conducting a comprehensive search across major open data repositories (e.g. OpenNeuro, TCIA, Synapse, ADNI, UK Biobank), peer-reviewed literature, and challenge websites. We included datasets that met the following criteria. (1) The primary imaging modality includes the brain (MRI, PET, CT, US, and SPECT. We do not consider EEG signals); and (2) public or institutional access is available with sufficient documentation that includes at least basic metadata (e.g., age, diagnosis, task labels); These criteria led to a curated list of 161 brain imaging datasets.

Next, we surveyed the literature on foundation models by compiling and filtering the models that are relevant to brain imaging from the three largest FM surveys ([8,18]). We curated 86 brain foundation models and analyzed in detail their experimental protocol. In particular, we have collected the datasets used to train and evaluate each of these models and the tasks for which each model was trained.

3.2 Dataset Annotation Protocol

Each dataset was manually reviewed and annotated with standardized metadata to allow filtering, comparison, and downstream use. We record the following:

- **Name:** Official name of the dataset.
- **Reference:** Citation(s) of original publications or technical reports.
- **Description:** Brief summary of the scope and purpose of the dataset.
- **Modalities/Sequences:** Imaging modalities or sequences included (e.g., MRI, PET, CT, SPECT, US).
- **Specific Pathologies:** Targeted diagnoses or conditions (e.g., Healthy, Parkinson's, Alzheimer's, Glioma, Metastasis, Pediatric disorders).
- **General Pathology Category:** Broader classification of diseases (e.g., neurodegeneration, cancer, vascular pathology, mental health condition and other neurological disorders).
- **Studies:** Number of imaging studies (that is, unique imaging sessions per subject, possibly containing multiple series).
- **Source URL:** Link to the official dataset access page.
- **Based On:** If the dataset is derived from or built on another dataset.
- **Anonymization:** Type of anonymization applied (e.g., defaced, skull stripped, deidentified, pseudonymized, fully anonymized).
- **Access Requirements:** Access control level (e.g., public, Data Use Agreement (DUA), Data Transfer Agreement (DTA), restricted).
- **Data Origin Countries:** Country or countries where data was collected.
- **Associated Projects:** Related research projects or consortiums (e.g. ADNI, HCP, UK Biobank).
- **Collection Year(s):** Time period during which data was collected.

Studies, Series, and Sequences In clinical imaging workflows, a study typically refers to a complete imaging session conducted for a patient, which may include several series. Each series comprises a set of image acquisitions performed with a consistent protocol or sequence of imaging parameters.

The fundamental units of a series are slices, individual 2D images that collectively form the dataset. Depending on how the data are structured, the resulting dataset can be organized in different dimensions: 2D datasets contain isolated slices or sections; 3D datasets are composed of volumetric data (stacks of slices); and 4D datasets involve a temporal or contrast-related extension of 3D volumes, capturing variations over time or acquisition parameters.

4 Results

In the following, we present our findings through the percentage and absolute numbers, which total is higher than the total count; it is worth noting that one dataset can cover multiple conditions, multiple modalities, etc. To explore the intersections, we refer to the detailed atlas linked in Sect. 1.

Table 1. Summary of datasets by imaging modality, protocol and tracer.

	MRI					
Protocol/Contrast	T1	T2	Flair	fMRI	T1Ce	DWI
# Datasets	128	68	60	56	43	26
# Studies	551309	350452	127028	387807	99707	46064
Protocol/Contrast	DTI	DSC	DCE	ASL		
# Datasets	17	10	5	5		
# Studies	65533	77757	743	4270		
	PET					
Tracer	AV45	FDG	PIB	Tau	FMISO	Beta	DASB
# Datasets	5	4	4	4	1	1	1
# Studies	5908	3846	4909	7602	423	3368	1100
Tracer	CUMI	Flumazenil	Raclopride	DTBZ	AV133	FE-PE2I
# Datasets	1	1	1	2	1	1
# Studies	1100	1100	298	981	683	100

	CT	SPECT	US
# Datasets	15	2	3
# Studies	27055	967	22335

Overview of Datasets by Modality. Our dataset atlas spans 151 brain datasets encompassing over 541k voumetric imaging studies, with MRI being the dominant modality both in terms of dataset count and study volume, Tab. 1). The most common protocols and contrasts are T1w, T2w, and BOLD/fMRI, while advanced sequences such as DTI, DSC, DCE, and ASL are less frequent. PET spans a wide range of tracers for neurodegeneration (AV45, PIB, Tau), metabolism (FDG, FMISO), psychiatry (DASB, CUMI), epilepsy (Flumazenil) and Parkinson's (Raclopride, DTBZ, AV133, FE-PE2I). CT is moderately represented (15 datasets) additional SPECT and ultrasound appear in only a few datasets.

In general, the field remains heavily dependent on MRI, with limited multimodal coverage (Perfusion) other modalities (PET, CT).

Overview of Datasets by Medical Condition. A broad spectrum of conditions is covered (Fig. 2). Healthy subjects make up the largest share, with 53 datasets and more than 392,000 studies, forming a strong baseline for control analyzes. Among diseases, Alzheimer's has the most studies (14 datasets, 38,625 studies), while mental health disorders such as schizophrenia and epilepsy account for 17 datasets (34,805 studies). Oncology is also well represented: glioma (27 datasets, 12,708 studies), GBM (21, 13,024), metastases (9, 9,503) and other cancers (18, 33,458). Stroke (14, 13,443) and Parkinson's (6, 4,430) add further variety.

Overall, the pathologies' coverage is very unbalanced. In particular, glioma has the most datasets, while Alzheimer's leads in study volume. When aggregated, cancer-related conditions dominate in both dataset count and imaging volume. In contrast, neurodegenerative disorders (particularly Parkinson's disease) remain comparatively underrepresented despite their clinical importance.

Overview of Datasets by Impact. We focus in the following on the three pathologies with the most datasets, cancer, Alzheimer (Fig. 3), and Parkinson diseases

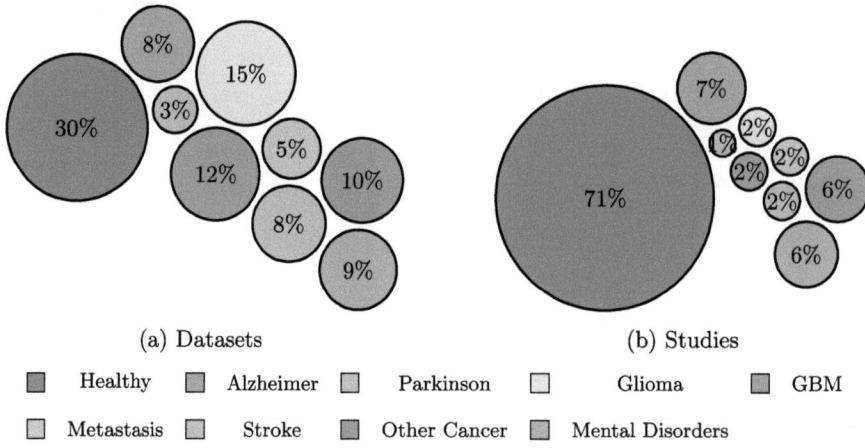

Fig. 2. The distribution of datasets and studies by medical condition

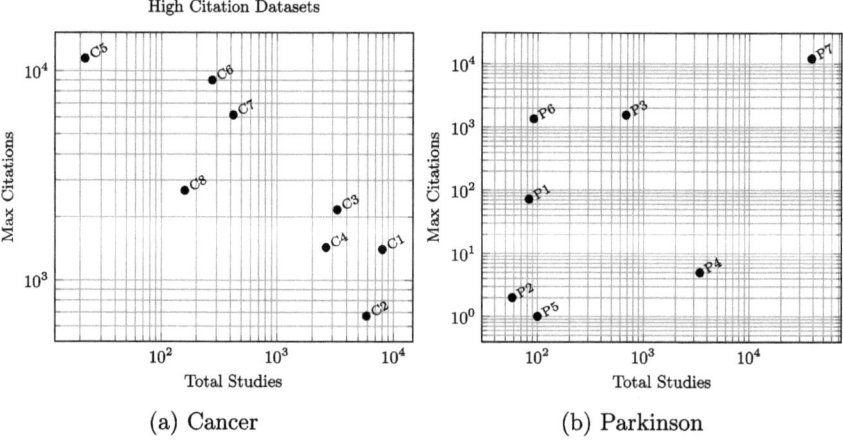

Fig. 3. Discrepancy Between Dataset Size and Impact: Brain imaging datasets for cancer and Parkinson disease, highlighting the four most cited datasets and the four datasets with the largest number of studies for each condition. **Cancer**: C1 NYUMets, C2 BraTS23 Glioma, C3 UPENN-GBM, C4 UCSC Xena Browser, C5 Curious, C6 CGGA, C7 ACRIN-FMISO 6684, C8 LGG-1p19qDeletion. **Parkinson's disease**: P1 INDI-Parkinsons, P2 OpenNeuro-ds004392, P3 PPMI, P4 ROSMAP, P5 OpenNeuro-ds005895, P6 OpenNeuro-ANT, P7 UKBiobank.

(Fig. 4). Across all datasets, we notice that the largest datasets (the largest number of studies) are not necessarily the most popular. The largest cancer dataset is NYUMets [12], while the most cited is the Curious dataset [20]. For both Parkinson and Alzheimer, the relevant dataset with the most citations and studies is the UKBioBank [19]. However, it is worth nothing that this is a commercial dataset and gathers a collection of dozens of cohorts covering many organs and pathologies. For Alzheimer, combining citations and studies in the 4 cohorts of OASIS [10,11,14,15] leads to a number of studies and citations of magnitude similar to UKBioBank. It should also be noted that some datasets with a large number of studies have very few citations (MIRIAD [13], BrainAge [4], Rosmap [1], or IvyGAP [16]).

There is a mismatch between the citation rate and size of datasets that hints that some important datasets are under-considered.

5 Discussion

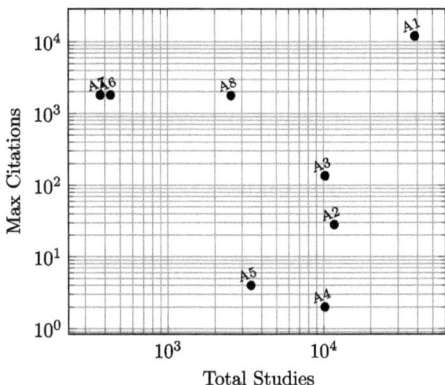

Fig. 4. Discrepancy Between Dataset Size and Impact: Brain imaging datasets for Alzheimer's disease, highlighting the four most cited datasets and the four datasets with the largest number of studies: A1 UKBioBank, A2 NACC, A3 iSTAGING, A4 BrainAge, A5 ROSMAP, A6 OASIS-1, A7 OASIS-2, A8 CBMFM

Our study identified the main patterns and limitations of the available brain imaging datasets that could be used for FM. In the following, we discuss and contrast our insights on brain datasets with our findings from the extended study of brain FMs [6]. We provide a detailed list of FMs online https://ballistic-purple-6f5.notion.site/182bf710ca5980d6b3a4fd3402960f0e.

Pathologies Coverage. The variety of pathologies represented in FMs differs considerably from what we observed by examining public datasets. Cancer is the most widely represented category in FMs, appearing in 34 models. Vascular diseases are represented in 10 models, just like neurodegeneration, while neurodevelopmental disorders appear in five models, and mental health conditions in three models. Conversely, our dataset analysis suggests a greater number of public studies focused on neurodegenerative conditions like Parkinson's and Alzheimer's compared to those concerning cancer.

Furthermore, our analysis highlights the disproportionate focus of current models on a small subset of diagnostic tasks, particularly brain tumor segmentation and Alzheimer's classification. Although important, this narrow scope overlooks numerous neurological and psychiatric disorders, including Parkinson's disease, ADHD, stroke subtypes, and rare pediatric conditions. Broader clinical relevance requires both dataset development and benchmarking efforts that expand beyond these well-trodden tasks.

In general, existing FM prioritize only a subset of available medical imaging datasets, are biased toward brain cancer research, and are less reflective of the actual prevalence of brain pathologies or available brain imaging datasets.

Biases. Numerous studies have shown that machine learning (ML) models in medical imaging can exhibit biases, often manifesting as disparities in diagnostic performance between demographic or clinical subgroups [5].

Only six foundational models factored in potential biases when constructing their datasets and models. NeuroBERT [3], for example, curated a balanced dataset reflecting various age groups (18 and older), social classes, and ethnicities (40% non-white). In addition, the authors developed age-specific embeddings to identify age-related irregularities. CLIP-T2Med [7] focused on gender bias during its evaluation. The researchers introduced a novel bias estimation metric to consider gender distribution, showing their method significantly reduces potential gender biases.

Although some biases can be mitigated in FM implementation, other biases inherent to the datasets are much more difficult to tackle. For example, there is a large imbalance in the country of origin of the public datasets collected. 76 datasets accounting for almost half of the brain studies (294.160) are from the USA, the two runners-up, far behind, are Netherlands (8 datasets and 129.054 studies) then China (10 datasets for a total of 37.320 studies).

Another disparity affecting FM generalization caused by the dataset is the discrepancy in the acquisition protocol. In more than half of the datasets (104), no information is provided about the machines or field strength (1.5T, 3T...) used in the collection. These acquisition settings can lead to very different image distributions. Similarly, the underutilization of raw k-space data poses a fundamental limitation: without access to pre-reconstruction data, many foundational tasks, such as learning generalizable representations, simulating artifacts, or improving sequence-specific fidelity, remain out of reach for FM researchers. Only 3 datasets provide raw signal data (RealNoiseMRI, FastMRI, and FastMRI+).

Meta-analysis of Best Models. In today's landscape, pinpointing the most effective architectures and models for specific tasks and datasets is challenging due to the abundance of datasets and baselines evaluated in various papers, coupled with the absence of standardized benchmarks and datasets in certain contexts. For example, in our study, of 40 FMs with anomaly segmentation, *Brain Tumor Segmentation (BraTS)* is the most popular benchmark, with only 15 models evaluating this dataset in their various iterations (BraTS2018, BraTS2019, BraTS2021, etc.). We illustrate the performance of these FMs in Table 2. The references of each models can be found on the atlases given the page constraints on the reference section. MoME [22], based on mixture of experts is the best performing model. This model uses a hierarchical gating network to combine the predictions of experts. In addition, it prevents the degeneration of each expert network and ensures their specialization using a curriculum learning strategy.

Access and Interoperability. Another critical challenge is dataset interoperability. The heterogeneity in imaging protocols, metadata formats, anonymization policies, and access terms complicates cross-cohort training and evaluation. For example, one of the largest datasets, UKBioBank, offers narrow commercial licenses, while 13 datasets use Creative Commons Non-Commercial licenses.

Meanwhile, for FM, 62 of the 86 models are sharing their source code and only four are restricting commercial usage.

Table 2. FMs evaluated on BraTS, with their dataset variants (BraTS19, BraTS21, ...) and reported DICE scores.

Model ↓	Version	DICE
Med3D	2013	55.41
Med-SAM-Brain	2023	61.85
Voco	2019	78.53
MoME+	2023	83.34
Eff-SAM	2018	83.40
PCRLv2	2018	85.60
MMC-Adapt	2021	86.90
LVM-Med	2021	90.40
Med-SA	2021	90.50
Genesis	2019	90.60
MedSegDiff-V2	2021	90.80
BrainSegFounder	2023	91.15
OBJ-SAM	2023	91.90
MoME	2021	92.10
MAE-Seg	Africa24	92.90

In general, without open, standardized schemas and preprocessing conventions, comparisons between models become difficult, and real-world deployment remains limited. Initiatives to harmonize the dataset schemas, such as common data elements, unified labeling protocols, and flexible input pipelines, could accelerate FM development and benchmarking.

Closing Words. We note a misalignment between the data scale and the impact. Our results show that the most cited datasets are not necessarily the most extensive, suggesting that accessibility, usability, and clinical relevance may play a larger role in adoption than volume alone. Promoting best practices around documentation, access, and openness will be as important as scaling up imaging archives. Taken together, our work illustrates that neuroimaging is not suffering from data scarcity, but from data fragmentation. FMs for brain imaging are expected to benefit substantially from improved dataset curation, broader task coverage, richer metadata, and lower technical barriers to adoption. Addressing these challenges is essential not only for technical advancement, but also for realizing the translational potential of FMs in neuroscience and clinical care.

6 Conclusion

Foundation Models (FMs) are reshaping the landscape of AI in medicine, but their application to brain imaging remains fragmented and narrowly focused. In this work, we present the most comprehensive atlas of brain imaging datasets to date, spanning diverse modalities, pathologies, and acquisition protocols.

To guide future research, we curated the largest atlas of brain imaging datasets to 151 datasets comprising more than 541k imaging studies—highlighting a rich but underutilized landscape of neuroimaging data. Our analysis identified a overwhelming dominance of structural MRI and a strong reliance on a few popular datasets, alongside critical blind spots that include underrepresented conditions like Parkinson's disease and overlooked data modalities such as advanced perfusion imaging or PET tracers.

Yet, beyond quantity, our findings underscore systemic issues that challenge scalability and reproducibility. Brain imaging datasets are highly heterogeneous in acquisition protocols, annotation formats, and metadata structures. This lack of standardization hinders the development of universal preprocessing pipelines and compromises cross-dataset model comparison, ultimately limiting FM generalizability. Moreover, the widespread absence of raw k-space MRI data restricts innovation in fundamental areas like image reconstruction and physics-informed modeling, slowing progress at the core of medical image analysis. To unlock the full potential of brain-specific FMs, we advocate for: (1) curated benchmarks with consistent protocols, (2) transparent reporting and documentation standards, (3) broader inclusion of underrepresented diseases and modalities, and (4) improved access to raw imaging data, particularly k-space.

We believe that by building on these recommendations, the research community can shift from fragmented efforts to robust, scalable, and clinically meaningful FMs for brain imaging.

Acknowledgement. This work was supported by the National Research Fund of Luxembourg (FNR), grant number C24/IS/18942843.

References

1. Bennett, D.A., Buchman, A.S., Boyle, P.A., Barnes, L.L., Wilson, R.S., Schneider, J.A.: Religious orders study and rush memory and aging project. J. Alzheimer's Disease: JAD **64**, S161 (2018). https://doi.org/10.3233/JAD-179939
2. Boudissa, S., Kanli, G., Perlo, D., Jaquet, T., Keunen, O.: Addressing artefacts in anatomical MR images: a k-space-based approach. In: IEEE International Symposium on Biomedical Imaging (2024). https://github.com/TransRad/MRArt.git
3. Wood, D., et al.: NeuroBERT: a self-supervised text-vision framework for automated brain abnormality detection (2024). https://doi.org/10.48550/arXiv.2405.02782
4. Feng, X., Lipton, Z.C., Yang, J., Small, S.A., Provenzano, F.A.: Estimating brain age based on a uniform healthy population with deep learning and structural magnetic resonance imaging. Neurobiol. Aging **91**, 15–25 (2020). https://doi.org/10.1016/J.NEUROBIOLAGING.2020.02.009
5. Ghamizi, S., Cordy, M., Papadakis, M., Le Traon, Y.: On evaluating adversarial robustness of chest X-Ray classification: pitfalls and best practices. In: The Thirty-Seventh AAAI Conference on Artificial Intelligence (AAAI-23)-SafeAI (2022)
6. Ghamizi, S., Kanli, G., Deng, Y., Perquin, M., Keunen, O.: Brain imaging foundation models, are we there yet? A systematic review of foundation models for brain imaging and biomedical research. arXiv preprint arXiv:2506.13306 (2025)
7. Xu, H., et al.: CLIP-T2MedI: fair text to medical image diffusion model with subgroup distribution aligned tuning (2025). https://doi.org/10.1117/12.3046450
8. He, Y., et al.: Foundation model for advancing healthcare: challenges, opportunities, and future directions (2024). https://doi.org/10.48550/arXiv.2404.03264
9. Kanli, G., et al.: Quantitative pre-clinical imaging of hypoxia and vascularity using MRI and pet. Methods Cell Biol. **191** (2025). https://doi.org/10.1016/BS.MCB.2024.10.016, https://pubmed.ncbi.nlm.nih.gov/39824561/
10. Koenig, L.N., et al.: Select atrophied regions in Alzheimer disease (sara): an improved volumetric model for identifying Alzheimer disease dementia. NeuroImage. Clin. **26** (2020). https://doi.org/10.1016/J.NICL.2020.102248, https://pubmed.ncbi.nlm.nih.gov/32334404/
11. LaMontagne, P.J., et al.: OASIS-3: longitudinal neuroimaging, clinical, and cognitive dataset for normal aging and Alzheimer disease (2019). https://doi.org/10.1101/2019.12.13.19014902, http://medrxiv.org/lookup/doi/10.1101/2019.12.13.19014902

12. Link, K.E., et al.: Longitudinal deep neural networks for assessing metastatic brain cancer on a large open benchmark. Nat. Commun. **15**, 1–10 (2024). https://doi.org/10.1038/s41467-024-52414-2, https://www.nature.com/articles/s41467-024-52414-2
13. Malone, I.B., et al.: MIRIAD–public release of a multiple time point Alzheimer's MR imaging dataset. NeuroImage **70**, 33–36 (2013). https://doi.org/10.1016/J.NEUROIMAGE.2012.12.044, https://pubmed.ncbi.nlm.nih.gov/23274184/
14. Marcus, D.S., Fotenos, A.F., Csernansky, J.G., Morris, J.C., Buckner, R.L.: Open access series of imaging studies: longitudinal MRI data in nondemented and demented older adults. Journal of cognitive neuroscience **22**, 2677–2684 (2010). https://doi.org/10.1162/JOCN.2009.21407, https://pubmed.ncbi.nlm.nih.gov/19929323/
15. Marcus, D.S., Wang, T.H., Parker, J., Csernansky, J.G., Morris, J.C., Buckner, R.L.: Open access series of imaging studies (oasis): cross-sectional MRI data in young, middle aged, nondemented, and demented older adults. J. Cogn. Neuroscience **19**, 1498–1507 (2007). https://doi.org/10.1162/JOCN.2007.19.9.1498, https://pubmed.ncbi.nlm.nih.gov/17714011/
16. Pati, S., et al.: Reproducibility analysis of multi-institutional paired expert annotations and radiomic features of the ivy glioblastoma atlas project (Ivy GAP) dataset. Med. Phys. **47**, 6039–6052 (2020). https://doi.org/10.1002/MP.14556, https://pubmed.ncbi.nlm.nih.gov/33118182/
17. Ranjbarzadeh, R., Caputo, A., Tirkolaee, E.B., Ghoushchi, S.J., Bendechache, M.: Brain tumor segmentation of MRI images: a comprehensive review on the application of artificial intelligence tools. Comput. Biol. Med. **152**, 106405 (2023). https://doi.org/10.1016/J.COMPBIOMED.2022.106405
18. Shi, C., Rezai, R., Yang, J., Dou, Q., Li, X.: A survey on trustworthiness in foundation models for medical image analysis (Towards) (2024). https://doi.org/10.48550/arXiv.2407.15851
19. Sudlow, C., et al.: UK biobank: an open access resource for identifying the causes of a wide range of complex diseases of middle and old age. PLoS Med. **12** (2015). https://doi.org/10.1371/JOURNAL.PMED.1001779, https://pubmed.ncbi.nlm.nih.gov/25826379/
20. Xiao, Y., et al.: Evaluation of MRI to ultrasound registration methods for brain shift correction: the curious2018 challenge. IEEE Trans. Med. Imaging **39**, 777–786 (2020). https://doi.org/10.1109/TMI.2019.2935060, https://pubmed.ncbi.nlm.nih.gov/31425023/
21. Yu, F., et al.: Evaluating progress in automatic chest X-Ray radiology report generation. Patterns **4**, 100802 (2023). https://doi.org/10.1016/j.patter.2023.100802
22. Zhang, X., et al.: MoME: a foundation model for brain lesion segmentation with mixture of modality experts (2024). https://doi.org/10.48550/arXiv.2405.10246

LGE Scar Quantification Using Foundation Models for Cardiac Disease Classification

Athira J. Jacob[1,2(✉)], Puneet Sharma[1], and Daniel Rueckert[2,3]

[1] Digital Technology and Innovation, Siemens Healthineers, Princeton, NJ, USA
[2] AI in Healthcare and Medicine, Klinikum rechts der Isar, Technical University of Munich, Munich, Germany
athira.jacob@siemens-healthineers.com
[3] Department of Computing, Imperial College London, London, UK

Abstract. Late Gadolinium Enhancement (LGE) cardiac MRI (CMRI) is the gold standard for assessing myocardial viability, but myocardial scar quantification is challenging due to inter-observer variability. While deep learning methods can achieve high agreement within a dataset, their generalization and clinical utility remain underexplored. In this work, we develop an LGE scar quantification model trained on pseudo-ground truth masks from expert-generated segment-wise labels. The dataset includes 159 and 53 patients for training and testing, respectively. To enhance performance, the model is initialized with a CMRI foundation model pre-trained on 36 million images. The model obtained an AUC of 0.96 and accuracy of 0.89 for LGE detection. The trained model is then applied to an independent dataset of 662 patients from a different institution, without any modification. Here, we demonstrate the utility of the calculated scar burden as a prognosticator for detecting myopathologies (normal, dilated, hypertrophic, ischemic), over and above other cine-based measurements, by improving AUCs by 1–6% for various disease classes. This study demonstrates that a scar segmentation model trained without expert-annotated pixel-wise ground truth masks can generate clinically meaningful scar burden measurements, offering valuable diagnostic insights.

Keywords: LGE scar segmentation · foundation model · cardiac disease classification · vision transformers

1 Introduction

Late Gadolinium Enhancement (LGE) imaging is the gold standard for assessing myocardial viability, providing clinical insights into scarred and fibrotic tissue. [11]. However, enhancement/scar analysis is challenging due to complex enhancement patterns, imaging artifacts (e.g. due to motion), scanner variations, and inter-observer inconsistencies in scar annotation, making it difficult to establish a precise and reliable pixel-wise ground truth (GT). On the other hand, binary labels indicating the presence or absence of enhancement are easier to create, though they are also subject to inter-user variability in borderline cases.

Traditional methods for LGE enhancement quantification rely on intensity thresholding techniques, such as the n-standard deviation (n-std) method and full-width at half-maximum (FWHM). While effective in some cases, these approaches are highly sensitive to imaging artifacts and variations in scan protocols [4,7]. Deep learning (DL)-based methods have been explored to address these limitations; however, most prior studies train models on limited datasets with small numbers of manually annotated LGE images, restricting their generalizability [14].

This work presents a method for LGE quantification and demonstrates its utility in classifying myopathologies. The model is trained using "pseudo-ground truth" (GT) created from segment-wise scar labels. The proposed method obtains an AUC of 0.96 and an accuracy of 0.89 for LGE detection in the same-center validation. Additionally, we demonstrate that our predicted scar burden (scar volume as a percentage of myocardial volume) improves myopathology classification in an independent external cohort, improving AUCs by 1–6% and accuracy by 3%, without requiring additional fine-tuning. Our results demonstrate a scalable LGE segmentation framework that enhances both segmentation reliability and clinical decision-making.

2 Related Work

LGE Detection. No consensus exists on the optimal quantitative method for assessing infarction in LGE images. Commonly used approaches include Full Width Half Maximum (FWHM) [1,8], and n-standard deviation (n-std) from remote myocardium. However, comparison studies [4,7] reveal substantial inter-technique variability, highlighting concerns regarding reliability and reproducibility. Numerous studies have investigated the use of DL techniques for scar detection from LGE [5,12,19,20]. Several of these, including [5,19,20], were conducted on the publicly available EMIDEC challenge dataset [13]. This dataset comprises LGE-MRI scans for 150 patients, approximately two-thirds of whom have pathological findings. Various methods achieved a normal/abnormal classification accuracy of up to 91% on the test set of 50 patients [14]. While challenge datasets [13,21] have been widely utilized, they present limitations that may hinder clinical translation, such as standardized image quality and the exclusion of slices without myocardium, which do not reflect real-world clinical data. Popescu et al. [17] used 155 LGE-MRI datasets, and 246 synthetic LGE datasets to train a model to segment scar and obtain scar burden, and acheieved a balanced accuracy of 75%. In [12], segmental information from diagnostic reports was used to learn presence/absence of LGE from LGE images. The method was developed on a cohort of 170 patients without any manual scar annotations (similar to our method), and reported an AUC of 0.875 for scar detection.

Disease Classification. Many studies have explored disease detection on the publicly available ACDC dataset [2] consisting of 150 patients, and achieved high accuracies of up to 0.96. However, this dataset did not include any LGE images and patients with ambiguous characteristics were excluded, potentially affecting performance. Paciorek et al. [16] train a DL model to distinguish normal/abnormal patients with 88% accuracy in a test set of 40 patients, however the specific types of diseases were not distinguished. In [10], the authors trained a model to classify patients into 4 groups (normal and 4 disease classes) based on cine images and obtained AUCs of 0.86 to 0.95 for the various classes.

3 Methods

3.1 Data

Center 1: Algorithm Development Dataset. The dataset consisted of 212 patients, divided into 159 patients (1516 images) for training and 53 patients (497 images) for testing, with 20% of the training set used for validation. Scans were acquired on 1.5T Siemens MAGNETOM Aera scanners using T1-weighted inversion-recovery sequences (TR/TE: 3.4/1.3 ms; slice thickness: 6 - 7 mm; in-plane resolution: 1.3 to 2.5 mm^2). Segment-wise binary scar labels (based on the AHA 17-segment model) were annotated by an expert reader (1 = scar present, 0 = no scar), giving 17 labels for each patient. LGE enhancement was observed in 62% of patients.

Center 2: Disease Classification Dataset. The dataset was an independent cohort of 662 patients (different continent and healthcare system). Scans were acquired on 1.5T Siemens MAGNETOM Avanto scanners using T1-weighted PSIR sequences (TR/TE: 2.4 /1.1ms; slice thickness: 8 mm; in-plane resolution: 1.3 to 2.6 mm^2), along with cine bSSFP in standard short and long views. Patients were classified into four groups: a) normal (NORM, 43%): no structural or functional abnormalities on cardiac MRI, b) dilated cardiomyopathy (DCM, 11%): left ventricular (LV) dilation and systolic dysfunction, c) hypertrophic cardiomyopathy (HCM, 13%): LV hypertrophy without abnormal loading conditions and d) ischemic heart disease (IHD, 33%): myocardial ischemia from coronary artery disease (CAD), often leading to infarction and scarring detectable on LGE imaging. The labels were assigned based on expert consensus review of images, reports, and records by two experienced cardiovascular radiologists.

3.2 Baseline Methods

We implemented the n-standard deviation (n-std) thresholding algorithm, defining scar as myocardial pixels with intensity above $\mu + n\sigma$, where μ and σ were computed over normal (non-scarred) myocardium. First, The myocardium was automatically segmented and divided into AHA segments using the anterior right ventricular insertion point (RVIP) [3]. For this, we used previously trained deep learning models for both myocardium segmentation and RVIP detection, offering a fast and reliable alternative to manual annotation. On independent test sets, these models achieved a Dice score of 0.89 for myocardium segmentation and a 3.1 mm localization error for RVIP detection—within the range of interobserver variability [22]. To estimate an upper bound on performance, normal myocardium regions were defined using expert-provided segment-wise labels (see Sect. 3.1, Center 1). Thresholds of $n = 2, 3$, and 4 were tested, with higher values yielding more conservative scar segmentation.

3.3 Psuedo-GT Generation on Algorithm Development Dataset

The proposed model was trained using pseudo-ground truth (pseudo-GT) masks, derived from the n-std thresholding method in combination with segment-wise clinical labels (Sect. 3.1). First, n-std method was run with n = 1.5 for maximum scar coverage. This method often overestimated scar, producing false positives (FP) in normal segments (Fig. 1a–d). To mitigate this, we manually refined the masks by removing FP clusters within clinically labeled normal segments, while preserving clusters in positive segments (shown in Fig. 1a–d). Specifically, connected components of the mask that were FP were deleted, while TPs were left as-is without any modifications. This approach retained the consistency and sensitivity of intensity-based delineation while improving specificity. We note that this method can also be potentially automated.

3.4 Proposed Model

The proposed model utilizes a U-Net transformer architecture [6], with a ViTS/8 [18] encoder. The encoder was initialized with a CMR foundation model pretrained in a self-supervised manner on 36 million cardiac MR images [9]. Ablation studies (Sect. 4.3) explore the effect of encoder architecture and pretraining. Input images were preprocessed by resampling to 1mm × 1mm and center cropped to 224 × 224 pixels. The model was trained using the Jaccard loss and a cosine annealing learning rate schedule (initial LR: 0.001), with geometric (rotation, translation) and elastic deformations applied for data augmentation. A multi-task learning strategy was employed, simultaneously predicting both the scar and myocardium masks, as multi-channel outputs for each image. Scar supervision was provided by the pseudo-GT masks described in Sect. 3.3, derived from segment-wise labels and intensity thresholding. Myocardium masks were obtained from the pretrained segmentation model detailed in Sect. 3.2. Final predictions were generated by ensembling three models trained with different random seeds.

Post-processing. Predicted masks were postprocessed as follows: a) Small isolated clusters—defined as connected components with area < 20 mm^2 and no other detected scar within a 5mm radius—were removed; (b) Slices above the basal extent of the left ventricle (LV) were excluded to avoid false positives near the LV outflow tract. The basal-most slice was identified using a pre-trained deep learning classifier applied to the corresponding cine stack. This location was then mapped to the LGE stack using slice position and orientation from the DICOM metadata. The full post-processing pipeline is illustrated in Fig. 1e.

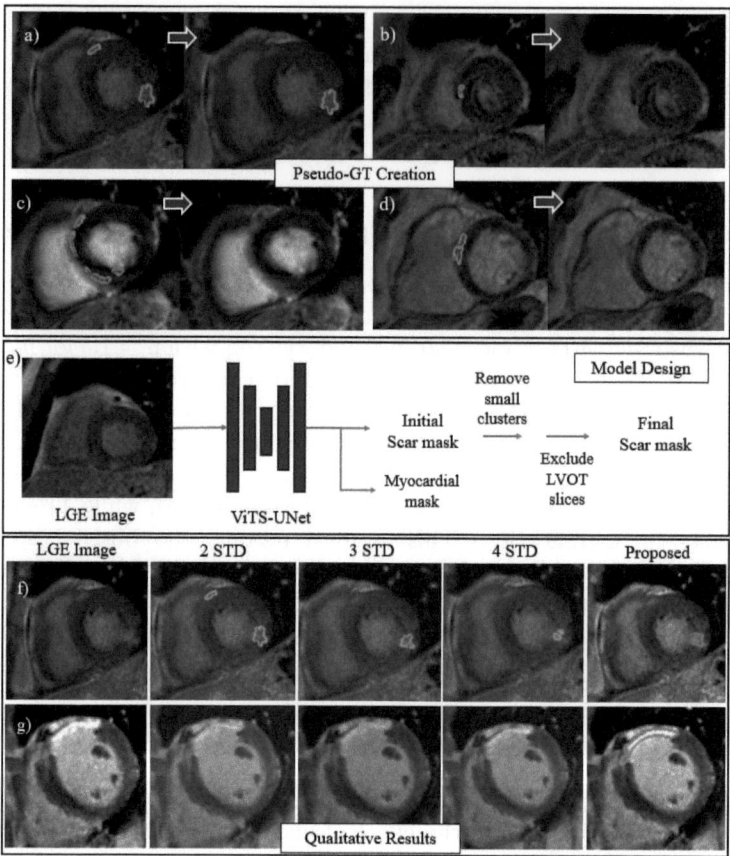

Fig. 1. (a–d) illustrate examples of initial masks from 1.5 std method that were refined to generate the psuedo-GT masks for training. False positive clusters were removed, while true positive clusters were kept as-is. (e) shows the pipeline used for inference. (f,g) Representative results on two patients from the same-center testing set.

3.5 Validation

Patient-Wise LGE Detection (Same-Center). Performance was evaluated using patient-level metrics for direct comparison with clinical ground truth, including accuracy, balanced accuracy, AUC, sensitivity, specificity, and a confusion matrix. AUC was assessed using scar burden, while patients with nonzero scar burden were classified as LGE-positive for other metrics.

Disease Classification (External). For each patient, LGE magnitude images were selected based on DICOM metadata. The dataset included two acquisition types: single-shot (acquired in a single cardiac cycle, without breath-hold) and segmented (acquired over multiple cycles, with breath-hold). Segmented images were preferred, when available, due to their higher spatial resolution for scar assessment. The trained scar quantification algorithm was applied without fine-tuning to estimate scar burden. This value was incorporated as an additional feature into an existing cine-based disease classification framework [10], which included features such as left ventricular ejection fraction, volumes, and strain. Two XGBoost classifiers were trained: (a) a cine-only model using 96 cine-image derived features and (b) a cine+LGE model using the quantified scar burden as an additional feature, giving 97 total features. The dataset was split into training(477), validation(54), and testing(171) sets, with 10-fold cross-validation. Performance was assessed using per-class AUC, weighted AUC, and overall accuracy on the test set. Feature importances are analyzed using SHAP analyses [15]. Note that the n-std algorithm, which requires a predefined normal myocardium region, could not be applied to this dataset due to the lack of such annotations.

Table 1. Results on same-center testing set (53 patients). The highest and second-highest values are highlighted in bold and underlined, respectively. Confusion matrices are presented as [True Negative, False Positive], [False Negative, True Positive].

Method	ROC-AUC	Accuracy	Balanced Accuracy	Confusion Matrix	Sensitivity	Specificity
2-std	0.76	0.68	0.61	[6 15][2 30]	0.94	0.29
3-std	0.76	0.70	0.68	[13 8][8 24]	0.75	0.62
4-std	0.71	0.66	0.70	[19 2][16 16]	0.50	**0.91**
Proposed	**0.96**	**0.89**	<u>0.86</u>	[16 6][0 31]	**1.00**	0.73

4 Results

4.1 Patient-Wise LGE Detection (Same-Center)

The performance of the proposed method and the n-std algorithm ($n = 2, 3, 4$) on the 53 test patients is summarized in Table 1. Increasing n in the thresholding algorithm results in a trade-off between sensitivity and specificity, with lower n values yielding higher sensitivity but lower specificity, vice-versa. Regardless, n-std method was seen to be biased towards FP. Figure 1 shows FPs due to RVIP (a), image artifacts (b), RV blood pool partial volume (PV) effects (c), and LVOT PV (d). The proposed model achieves the highest AUC and accuracy metrics. Representative scar segmentation results is shown in Fig. 1f-g.

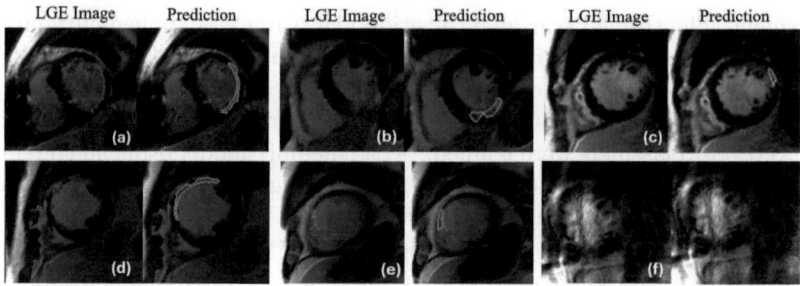

Fig. 2. Results on images from six patients on the external dataset. (a–d) True positive results, (e) Partial true positive, (f) False positive due to image artifact

4.2 Disease Classification (External)

The results for the disease classification dataset are shown in Table 2. Incorporating scar burden as a feature enhances classification performance across all classes, with AUC improvements of 1% for DCM, 2% for NORM and HCM, and 6% for IHD. The most significant gain is observed in IHD, aligning with its characteristic LGE presentation. SHAP analysis of the XGB model (Fig. 3) further emphasizes this, showing that LGE scar burden is the most influential feature for distinguishing IHD while also contributing to NORM and DCM classification, consistent with clinical protocols. Examples of predicted scar on this dataset are shown in Fig. 2.

Table 2. Results on disease classification in external dataset (n = 131 patients)

Model	ROC-AUC				Weighted AUC	Accuracy
	NORM	DCM	HCM	IHD		
Cine-only	0.95	0.87	0.93	0.83	0.90	0.76
Cine+LGE	**0.97**	**0.88**	**0.95**	**0.89**	**0.93**	**0.79**

4.3 Ablation Study Encoder Effect

The effect of encoder choice on performance was assessed by comparing three encoders: a) the CMR foundation model (ViTS) [9], b) a convolutional model trained on the related task of LGE myocardial segmentation (Dense121) (Sect. 3.2), and c) a ViTS model pre-trained on ImageNet. The CMR-trained ViTS achieved the best performance (Table 3), likely due to its ability to capture domain-specific cardiac features from large-scale data. The ImageNet-pretrained ViTS model achieved intermediate ROC-AUC but the lowest accuracy metrics, likely due to the domain gap. The convolutional encoder trained on the related task of myocardium segmentation achieved intermediate accuracy metrics but the lowest AUC, and may have been constrained by its task-specific training.

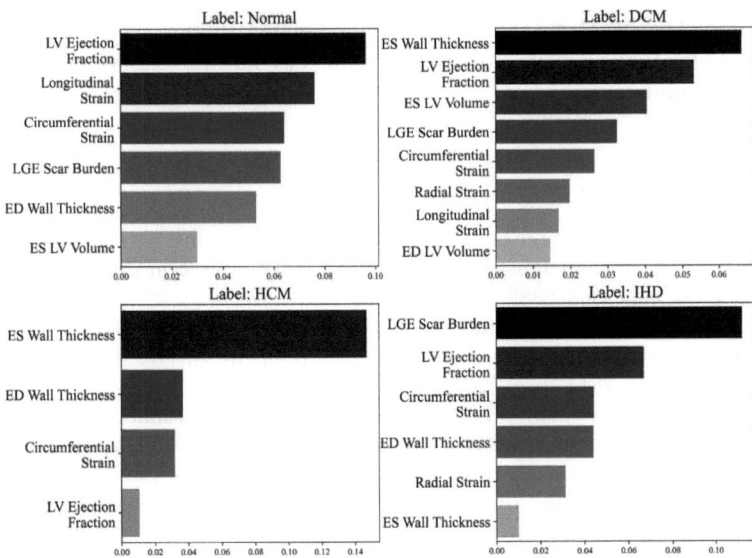

Fig. 3. Feature attributions for various disease classes using SHAP analyses.

Table 3. Ablation study: Effect of different encoders on performance. n = 53 patients

Encoder Type	ROC-AUC	Accuracy	Balanced Accuracy
CMR Foundation Model (ViTS/8)	**0.96**	**0.89**	**0.86**
Convolutional encoder (Dense121) pretrained on related task	0.82	0.81	0.83
ViTS/8 pretrained on ImageNet	0.89	0.75	0.71

5 Conclusions

This study presents a deep learning–based method for LGE scar quantification that demonstrates strong performance, cross-institution generalizability, and clinical utility in downstream disease classification tasks. Notably, the approach does not rely on expert-delineated pixel-wise ground truth masks. However, limitations include the use of single-vendor datasets, the absence of expert-annotated LGE masks for model comparison, and the lack of LGE ground truth labels in the disease classification cohort. Despite these constraints, the derived scar burden proved clinically meaningful, improving disease detection metrics and contributing to model interpretability through feature attribution analyses. Overall, the proposed method offers a scalable and annotation-efficient solution for generating robust LGE scar burden estimates, mitigating the challenges of manual labeling and inter-observer variability.

Disclosure of Interests. The authors AJJ and PS are employees of Siemens Healthineers.

Disclaimer. The concepts and information presented in this paper are based on research results that are not commercially available. Future commercial availability cannot be guaranteed.

References

1. Amado, L.C., et al.: Accurate and objective infarct sizing by contrast-enhanced magnetic resonance imaging in a canine myocardial infarction model. J. Am. Coll. Cardiol. **44**(12), 2383–2389 (2004)
2. Bernard, O., et al.: Deep learning techniques for automatic MRI cardiac multi-structures segmentation and diagnosis: is the problem solved? IEEE Trans. Med. Imaging **37**(11), 2514–2525 (2018)
3. Cerqueira, M., Weissman, N., Dilsizian, V., Jacobs, A.: Standardized myocardial segmentation and nomenclature for tomographic imaging of the heart: a statement for healthcare professionals from the cardiac imaging committee of the council on clinical cardiology of the American heart association. Circulation **105**(4), 539–542 (2002)
4. Flett, A.S., et al.: Evaluation of techniques for the quantification of myocardial scar of differing etiology using cardiac magnetic resonance. JACC: Cardiovasc. Imaging **4**(2), 150–156 (2011)
5. Girum, K.B., Skandarani, Y., Hussain, R., Grayeli, A.B., Créhange, G., Lalande, A.: Automatic myocardial infarction evaluation from delayed-enhancement cardiac MRI using deep convolutional networks. In: Statistical Atlases and Computational Models of the Heart. M&Ms and EMIDEC Challenges: 11th International Workshop, STACOM 2020, Held in Conjunction with MICCAI 2020, Lima, Peru, October 4, 2020, Revised Selected Papers 11, pp. 378–384. Springer (2021)
6. Hatamizadeh, A., et al.: UNETR: transformers for 3D medical image segmentation. In: Proceedings of the IEEE/CVF Winter Conference on Applications of Computer Vision, pp. 574–584 (2022)
7. Heiberg, E., et al.: Infarct quantification with cardiovascular magnetic resonance using "standard deviation from remote" is unreliable: validation in multi-centre multi-vendor data. J. Cardiovasc. Magn. Reson. **24**(1), 53 (2022)
8. Hsu, L.Y., Natanzon, A., Kellman, P., Hirsch, G.A., Aletras, A.H., Arai, A.E.: Quantitative myocardial infarction on delayed enhancement MRI. part i: Animal validation of an automated feature analysis and combined thresholding infarct sizing algorithm. J. Magn. Reson. Imaging: Official J. Int. Soc. Magn. Reson. Med. **23**(3), 298–308 (2006)
9. Jacob, A.J., Borgohain, I., Chitiboi, T., Sharma, P., Comaniciu, D., Rueckert, D.: Towards a vision foundation model for comprehensive assessment of cardiac MRI. arXiv preprint arXiv:2410.01665 (2024)
10. Jacob, A.J., et al.: Deep-learning-based disease classification in patients undergoing cine cardiac MRI. J. Magn. Reson. Imaging (2024)
11. Jenista, E.R., et al.: Revisiting how we perform late gadolinium enhancement CMR: insights gleaned over 25 years of clinical practice. J. Cardiovasc. Magn. Reson. **25**(1), 18 (2023)
12. Kim, Y.C., Chung, Y., Choe, Y.H.: Deep learning for classification of late gadolinium enhancement lesions based on the 16-segment left ventricular model. Phys. Med. **117**, 103193 (2024)

13. Lalande, A., et al.: Emidec: a database usable for the automatic evaluation of myocardial infarction from delayed-enhancement cardiac MRI. Data **5**(4), 89 (2020)
14. Lalande, A., et al.: Deep learning methods for automatic evaluation of delayed enhancement-MRI. The results of the emidec challenge. Med. Image Anal. **79**, 102428 (2022)
15. Lundberg, S.M., Lee, S.I.: A unified approach to interpreting model predictions. In: Guyon, I., Luxburg, U.V., Bengio, S., Wallach, H., Fergus, R., Vishwanathan, S., Garnett, R. (eds.) Advances in Neural Information Processing Systems 30, pp. 4765–4774. Curran Associates, Inc. (2017). http://papers.nips.cc/paper/7062-a-unified-approach-to-interpreting-model-predictions.pdf
16. Paciorek, A.M., et al.: Automated assessment of cardiac pathologies on cardiac MRI using t1-mapping and late gadolinium phase sensitive inversion recovery sequences with deep learning. BMC Med. Imaging **24**(1), 43 (2024)
17. Popescu, D.M., et al.: Anatomically informed deep learning on contrast-enhanced cardiac magnetic resonance imaging for scar segmentation and clinical feature extraction. Cardiovasc. Digit. Health J. **3**(1), 2–13 (2022)
18. Touvron, H., Cord, M., Douze, M., Massa, F., Sablayrolles, A., Jégou, H.: Training data-efficient image transformers & distillation through attention. In: International Conference on Machine Learning, pp. 10347–10357. PMLR (2021)
19. Yang, S., Wang, X.: A hybrid network for automatic myocardial infarction segmentation in delayed enhancement-MRI. In: Statistical Atlases and Computational Models of the Heart. M&Ms and EMIDEC Challenges: 11th International Workshop, STACOM 2020, Held in Conjunction with MICCAI 2020, Lima, Peru, October 4, 2020, Revised Selected Papers 11, pp. 351–358. Springer (2021)
20. Zhang, Y.: Cascaded convolutional neural network for automatic myocardial infarction segmentation from delayed-enhancement cardiac MRI. In: Statistical Atlases and Computational Models of the Heart. M&Ms and EMIDEC Challenges: 11th International Workshop, STACOM 2020, Held in Conjunction with MICCAI 2020, Lima, Peru, October 4, 2020, Revised Selected Papers 11, pp. 328–333. Springer (2021)
21. Zhuang, X.: Multivariate mixture model for myocardial segmentation combining multi-source images. IEEE Trans. Pattern Anal. Mach. Intell. **41**(12), 2933–2946 (2018)
22. Zhuang, X., et al.: Cardiac segmentation on late gadolinium enhancement MRI: a benchmark study from multi-sequence cardiac MR segmentation challenge. Med. Image Anal. **81**, 102528 (2022)

Beyond Broad Applications: Can Pathology Foundation Models Adapt to Hematopathology?

Chaima Ben Rabah(✉) and Ahmed Serag

AI Innovation Lab, Weill Cornell Medicine – Qatar, Doha, Qatar
chb4036@qatar-med.cornell.edu

Abstract. Recent advances in foundation models offer promising, scalable solutions for digital pathology, especially for classification tasks with limited data. However, most studies treat these models as static feature extractors, overlooking their full potential in real-world scenarios. Meanwhile, accurate blood cell morphology analysis is vital for diagnosing disorders like anemia and leukemia, particularly in resource-limited settings. In this study, we systematically evaluate the robustness and adaptability of four pathology foundation models specifically Gigapath, PLIP, DinoBloom, and PhikonV2, across five hematology datasets with diverse sizes, class distributions, and clinical contexts. We benchmark each foundation model under different training paradigms and compare them against standard deep learning architectures. Our results reveal that model rankings are highly dependent on the size of the training data and model configuration, with foundation models often underperforming or matching simpler models when not fine-tuned appropriately. These findings emphasize the need for robust, multi-regime benchmarking to fairly assess foundation models in pathology, especially in domain-specific, low-data applications like hematology.

Keywords: Foundation Models · Hematology · Digital Pathology · Fine-Tuning · Whole-Slide Images · Vision Transformers · Benchmarking

1 Introduction

Pathologists around the world face mounting pressures as demand increasingly outweighs supply, creating critical shortages in healthcare systems [17,20]. The modern pathologist's role has expanded beyond traditional microscopic examination to encompass complex molecular assays and genetic testing, further straining this limited workforce. Simultaneously, the digitization of pathology slides has created unprecedented opportunities for computational approaches, with artificial intelligence (AI) emerging as a promising solution to these challenges [11,19]. Foundation models (FM), large-scale AI systems designed to adapt across diverse applications, are now transforming computational pathology through their scalable architectures and multimodal integration capabilities, with particularly significant implications for specialized fields like hematopathology [7,9].

The evolution of AI in pathology has progressed from narrow task-specific tools to sophisticated FMs capable of addressing the unique challenges of pathology data. These challenges range from processing gigapixel whole-slide images (WSIs) to analyzing single-cell morphology, while enabling knowledge transfer across institutions and disease domains. Unlike conventional AI approaches that require extensive task-specific training, FMs can adapt to applications they weren't specifically designed for, offering remarkable flexibility in clinical and research settings. This adaptability is particularly valuable in hematopathology, where morphological subtleties and rare cellular events demand both contextual understanding and precise cellular analysis.

Microsoft's GigaPath [22] represents a paradigm shift in computational pathology with its innovative "LongNet" architecture, which processes entire WSIs as continuous sequences rather than fragmented patches. Trained on more than 170,000 slides from 28 US cancer centers, GigaPath's 2024 release included hematology-specific benchmarks that achieved 94% AUC in predicting TP53 mutations from acute myeloid leukemia slides, a 12% improvement over previous patch-based methods. The model's global context modeling capability has proven especially valuable for capturing rare events like circulating tumor cells in peripheral blood smears, a task where traditional convolutional neural networks (CNNs) often fail due to their limited receptive fields.

For multimodal integration, PLIP (Pathology Language and Image Pre-training) [13] provides an accessible framework through its HuggingFace implementation. It was trained using billion of pathology image–text pairs from Twitter and other entities including LAION. Its lightweight design facilitates deployment on edge devices in resource-limited hematology clinics, addressing practical implementation concerns.

Phikon-v2 [10] addresses scalability challenges through self-supervised pre-training on 460 million pathology tiles. A key distinctive feature that make Phikon-v2 notable is that it was trained to encode granular morphological features such as nuclear shape, tissue texture, and cellular patterns which are critical in hematopathology and other fine-detailed domains.

While general pathology FMs offer broad tissue analysis capabilities, DinoBloom [14] stands apart as the first specialized FM for hematology, representing a significant advancement in single-cell analysis. Unlike general pathology FMs that struggle with cellular-level resolution, DinoBloom was purposefully designed for hematologic applications and trained on 380,000 white blood cell images across diverse preparation methods. The 2025 extension of DinoBloom to minimal residual disease (MRD) detection in acute lymphoblastic leukemia further highlights its specialized advantage, demonstrating 0.1% sensitivity that surpasses both conventional flow cytometry thresholds and the detection limits of general pathology FMs when applied to hematologic specimens.

However, despite the potential of these FMs, their application in classification tasks within hematopathology remains underexplored. A key challenge lies in their limited adaptation to the domain-specific nuances of hematopathological imagery. FMs are typically trained on broad, heterogeneous datasets, which may not capture the fine-grained morphological features critical for accurate diagnosis in this field. This raises concerns about their reliability and clinical relevance when deployed in specialized settings without further tuning or domain-aware evaluation.

In this research, we investigate the adaptability of pathology FMs to hematopathology specific classification tasks. By leveraging pre-trained FMs and fine-tuning them on specialized hematopathology datasets, this paper seeks to fill the gap by evaluating several FMs through rigorous external testing across multiple datasets, providing a framework for responsible clinical translation. The results highlight both the remarkable potential and practical limitations of FMs in specialized pathology domains, offering insights into their optimal implementation for improving diagnostic accuracy, reproducibility, and efficiency in hematopathology practice.

2 Experimental Setup and Methods

This study evaluated the performance of four FMs (GigaPath, PLIP, PhikonV2, and Dinobloom) across different experimental settings and datasets in hematopathology classification tasks. The models were trained and evaluated using 5-fold cross-validation, with performance assessed using three key metrics: accuracy, F1-score, and the area under the curve (AUC). These results were then compared against the performance of three state-of-the-art deep learning (DL) models: MobileNetV3 [15], ResNet50 [12], and the Vision Transformer (ViT) [8] which has become a standard architecture due to its excellent performance on many computer vision benchmarks [4,16]. Each model was evaluated under two distinct training paradigms: (1) trained from scratch with randomly initialized weights, and (2) used as a feature extractor by loading pretrained weights from ImageNet [18] and fine-tuning only the final classification layer while keeping all other layers frozen. This dual setting allows for a comprehensive assessment of both model capacity and transfer learning effectiveness in the context of limited medical imaging data.

2.1 Pipeline for the Classification Task

We evaluated the performance of the FMs across three distinct experimental settings:

- **Feature Extraction without Fine-Tuning (Setting 1):** The FMs were employed solely as fixed feature extractors, without any additional training or fine-tuning on the target dataset. The embeddings generated by the pre-trained FMs served as input features for downstream classification.
- **Full Data Fine-Tuning of the Last Layer (Setting 2):** We fine-tuned only the final classification layer of the FMs using the entire available training set, except for a reserved test set comprising 100 images per class. This setting aims to assess the model's adaptability when trained on a comprehensive dataset while preserving a balanced and unseen evaluation subset.

- **Partial Data Fine-Tuning of the Last Layer (Setting 3):** Similar to setting 2, we fine-tuned only the last layer; however, the training data was reduced to half of the total available samples, again holding out 100 images per class for testing. This condition simulates limited data availability and evaluates the robustness of fine-tuning under constrained data regimes.

2.2 Evaluation Datasets

In this study, we comprehensively evaluate binary and multi-class classification performance across five diverse digital hematopathology datasets, which include both multi-cell and single-cell images, to ensure robust and generalizable conclusions in computational pathology. ALL image dataset (D1) [3] includes 3,256 peripheral blood smear images from 89 patients at Taleqani Hospital (Iran), categorized into benign hematogones and malignant lymphoblast subtypes, captured at 100x magnification (224 × 224 JPG). The WBC dataset (D2) [23] comprises 300 white blood cell images (120 × 120 pixels) from Jiangxi Tecom, China, stained with rapid hematology reagent and showing yellowish backgrounds. The ALL-IDB2 dataset (D3) [1] contains 280 cropped images focusing on individual cells (normal vs malignant) and is widely used for leukemia-related algorithm development. The PBS Barcelona (D4) [2] offers 17,092 expert-annotated images of normal blood cells in eight classes, acquired at the Hospital Clinic of Barcelona using a CellaVision DM96 analyzer (360 × 363 JPG). Finally, the Bodzas dataset (D5) [6] includes 16,027 high-resolution annotated images of 9 white blood cell types, derived from 12,986 samples of 78 patients (leukemic and non-leukemic), collected between 2020 and 2022 with a resolution of 42 pixels/μm. Table 1 provides a detailed overview of these datasets.

Table 1. Summary of the datasets used in this study.

ID	Images	Classes	Class Names
D1	3,256	4	Benign, Early Pre-B, Pre-B, Pro-B ALL
D2	300	4	Neutrophil, Lymphocyte, Monocyte, Eosinophil
D3	260	2	Normal, Malignant
D4	17,101	8	Neutrophils, Eosinophils, Basophils, Lymphocytes, Monocytes, Immature granulocytes, Erythroblasts, Platelets
D5	16,037	9	Neutrophil segments, Neutrophil bands, Eosinophils, Basophils, Lymphocytes, Monocytes, Normoblasts, Myeloblasts

3 Results and Discussion

3.1 Evaluation

Initially, we evaluated four FMs for histopathology using a zero-shot approach (Setting 1), wherein the models generated embeddings that were directly used for classification without any task-specific fine-tuning.

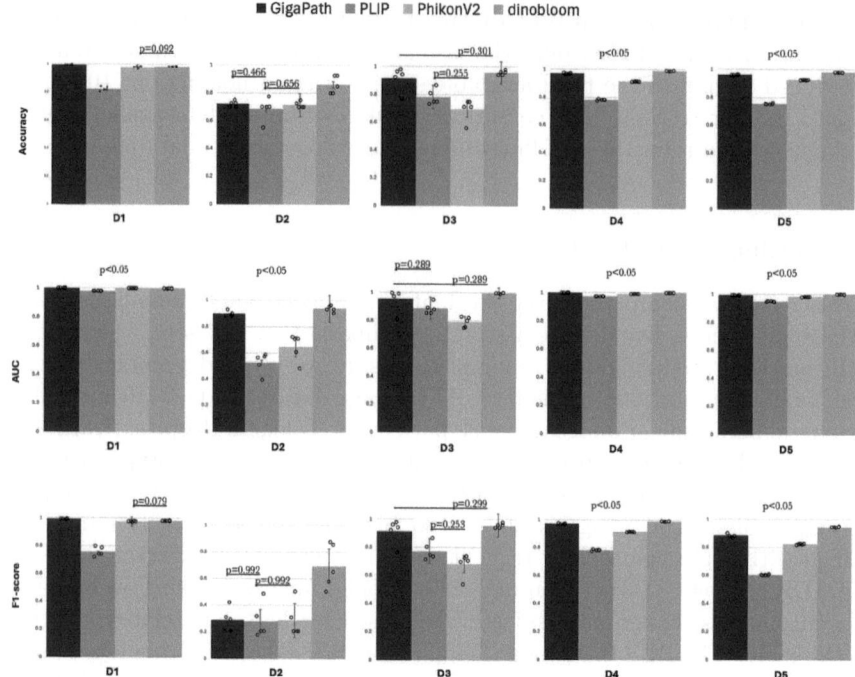

Fig. 1. Performance comparison of GigaPath, PLIP, PhikonV2 and dinobloom across five datasets using a zero-shot approach. Bar plots illustrate accuracy, F1-score, and AUC for each model on each dataset. The error bars show the standard error across 5 independent experiments. The p-values indicate the significance level using paired t-tests with FDR correction

Experiments were conducted on diagnostic images from five distinct datasets to assess model generalizability. Figure 1 presents the accuracy, F1-score, and AUC for each FM across all datasets.

Dinobloom demonstrates the best performance by achieving the highest accuracy, AUC and F1-score across D1, D2, D3 and D4 compared with the general pathology FMs. On the D1, we noted the performance was not significantly different from GigaPath. We also found that the performance of all models has decreased on D2 with dinobloom achieving substantial improvement against the other FMs.

Figure 2 presents t-SNE (t-Distributed Stochastic Neighbor Embedding) [21] visualizations of the image embeddings generated by the four FMs across the five datasets. These plots offer qualitative support for the quantitative performance metrics reported in Fig. 1. In particular, embeddings from Dataset D2 consistently show less clear class separation across all models, highlighting the difficulty in distinguishing its classes. Across all datasets, DinoBloom consistently produces embeddings with clearer cluster separation, characterized by distinct and well-formed clusters. This comparative visualization highlights DinoBloom's ability to generate more discriminative representations than the other models, regardless of the dataset.

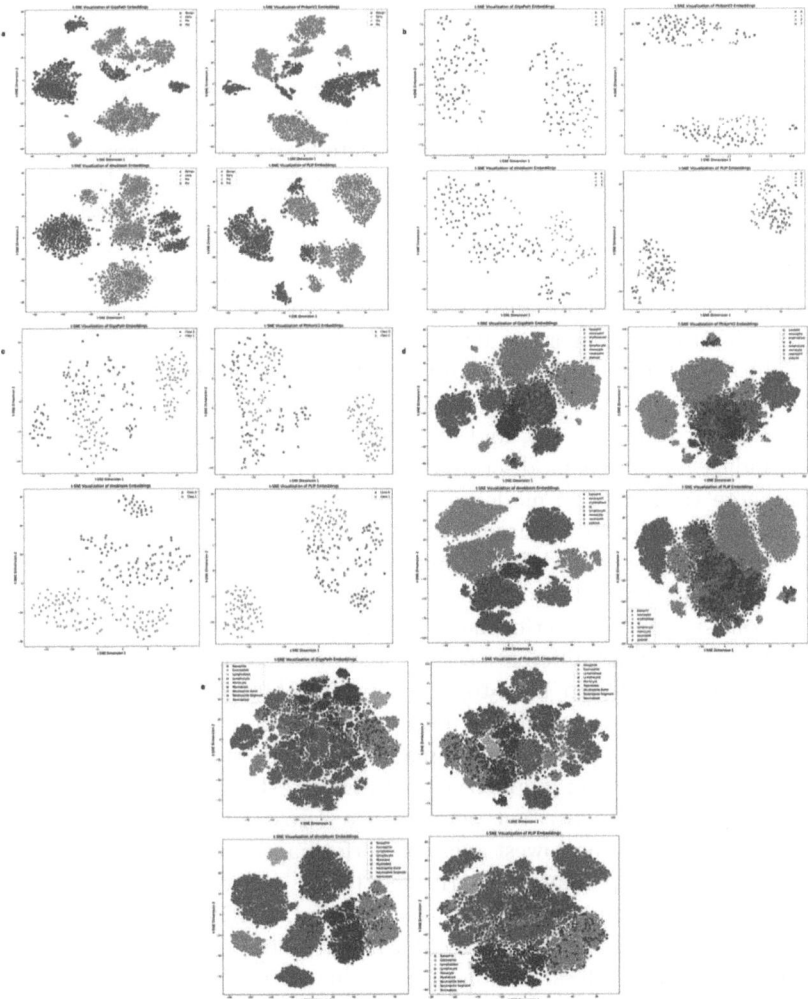

Fig. 2. t-SNE visualizations of embeddings from GigaPath, PLIP, PhikonV2 and Dinobloom across datasets D1-D5 (a-e). Each plot displays two-dimensional projections of high-dimensional feature spaces, where each point represents an image and colors indicate different classes.

3.2 Foundation Models Under Different Settings

To study the impact of fine-tuning on the performance of the FMs, we conducted several comparison under the settings described in Sect. 2.1. Due to dataset size limitations, only datasets D1 and D4 were included in this evaluation. Other datasets lacked sufficient image samples per class to reliably support the described experimental protocols, particularly the fixed-size test splits.

In Setting 1 with the D1 dataset, GigaPath demonstrated exceptional performance with an accuracy of 0.9942 ($\pm$ 0.0017), F1-score of 0.9937 ($\pm$ 0.0019), and near-perfect AUC of 0.9999 ($\pm$ 0.0001). This performance can be attributed to its innovative "LongNet" architecture, which processes entire whole-slide images as continuous sequences rather than fragmented patches. This approach allows GigaPath to capture global contextual information that might be missed by models with more limited receptive fields. Dinobloom followed closely with an accuracy of 0.9834 ($\pm$ 0.0023), F1-score of 0.9826 ($\pm$ 0.0026), and AUC of 0.9986 ($\pm$ 0.0002). In Setting 2 with the D1 dataset, all models demonstrated improved consistency in their performance. Dinobloom achieved the highest accuracy at 0.99 ($\pm$ 0.006) and F1-score at 0.990 ($\pm$ 0.006), with a perfect AUC of 1.000 ($\pm$ 0.000). GigaPath, PhikonV2, and PLIP followed with accuracies of 0.982 ($\pm$ 0.007), 0.984 ($\pm$ 0.005), and 0.960 ($\pm$ 0.025), respectively. The F1-scores mirrored these results closely, and all models achieved AUC values of 0.999 or higher, indicating excellent discriminative ability. Setting 3 with D1 revealed slightly more variability in model performance with GigaPath maintaining the strongest results with an accuracy of 0.966 ($\pm$ 0.023) and F1-score of 0.965 ($\pm$ 0.024). Dataset D4 revealed Dinobloom as the top performer with an accuracy and F1-score surpassing 0.98 and perfect AUC.

The consistently high performance of Dinobloom, pretrained on hematology data exclusively, across different settings and datasets demonstrates the potential of specialized FMs in hematopathology. Dinobloom appears to provide a significant advantage in hematopathology classification tasks. This aligns with previous findings that FMs with domain-specific pretraining can outperform more general models when applied to specialized tasks [5].

PhikonV2 showed consistent performance across all settings and datasets, while PLIP achieved the lowest relative performance among the four models. This performance gap may be attributed to its design focus on multimodal integration rather than pure image classification. PLIP's architecture, which links cytomorphological features to clinical text entries, may not be optimally aligned with the pure classification tasks evaluated in this study. Nevertheless, its improved performance in Settings 2 and 3 suggests that PLIP can adapt effectively to specific hematopathology tasks when provided with appropriate experimental conditions. Moreover, the consistently low performance across all models on certain datasets (D2 and D3) highlights the critical impact of dataset size on model effectiveness.

3.3 DL-Based Prediction Performance

Figure 3 reveals a comprehensive performance comparison of three DL models (MobileNetV3, ResNet50, and ViT) across the five datasets (D1-D5) using Accuracy, F1-score, and AUC metrics, demonstrating that datasets D1, D4, and D5 consistently yield higher performance while D2 and D3 present significant challenges, particularly for F1-score.

Fine-tuning substantially improves performance across all models and datasets with MobileNetV3 showing the most dramatic improvements (especially

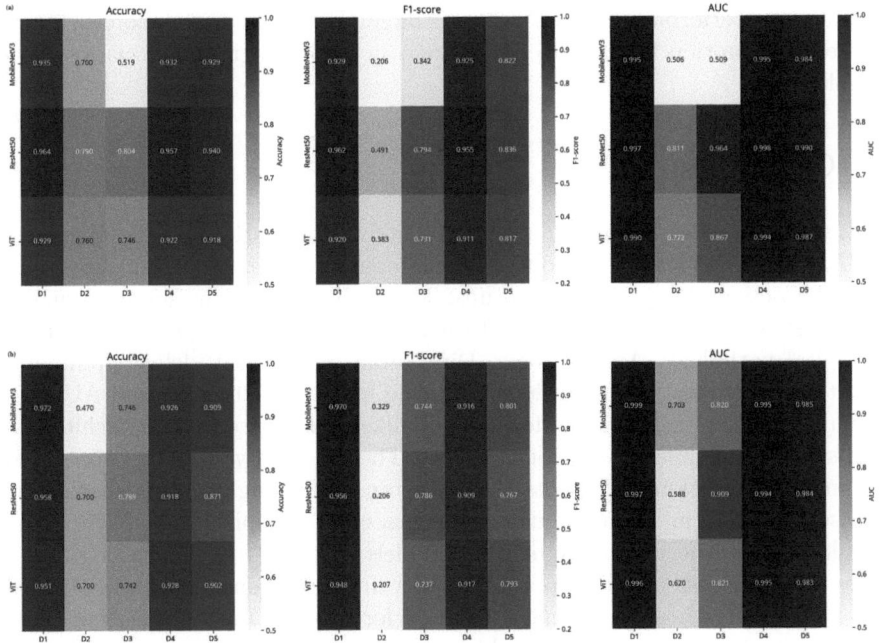

Fig. 3. Performance of the DL models (a) without fine-tuning (b) with fine-tuning.

Table 2. Model performance across three settings for Data1 and Data4 using Accuracy, F1-score, and AUC (mean ± std). Best values per setting and metric are in bold.

Setting	Model	D1			D4		
		Accuracy	F1-score	AUC	Accuracy	F1-score	AUC
Setting 1	GigaPath	**0.9942 ± 0.0017**	**0.9937 ± 0.0019**	**0.9999 ± 0.0001**	0.970 ± 0.002	0.969 ± 0.003	0.999 ± 0.000
	PLIP	0.8227 ± 0.0157	0.7568 ± 0.0331	0.9794 ± 0.0008	0.781 ± 0.005	0.721 ± 0.009	0.975 ± 0.000
	PhikonV2	0.9782 ± 0.0033	0.9767 ± 0.0032	0.9992 ± 0.0001	0.913 ± 0.002	0.900 ± 0.002	0.992 ± 0.000
	Dinobloom	0.9834 ± 0.0023	0.9826 ± 0.0026	0.9986 ± 0.0002	**0.989 ± 0.001**	**0.990 ± 0.001**	**1.000 ± 0.000**
Setting 2	GigaPath	0.982 ± 0.007	0.981 ± 0.007	**1.000 ± 0.000**	0.9862 ± 0.0046	0.9863 ± 0.0046	0.9998 ± 0.0002
	PLIP	0.960 ± 0.025	0.959 ± 0.025	0.999 ± 0.000	0.9757 ± 0.0077	0.9758 ± 0.0076	0.9995 ± 0.0003
	PhikonV2	0.984 ± 0.005	0.984 ± 0.005	0.999 ± 0.000	0.9867 ± 0.0037	0.9868 ± 0.0037	**0.9998 ± 0.0001**
	Dinobloom	**0.990 ± 0.006**	**0.990 ± 0.006**	**1.000 ± 0.000**	**0.9908 ± 0.0036**	**0.9908 ± 0.0036**	0.9998 ± 0.0001
Setting 3	GigaPath	0.966 ± 0.023	0.965 ± 0.024	**1.000 ± 0.001**	0.988 ± 0.003	0.988 ± 0.003	**1.000 ± 0.000**
	PLIP	0.919 ± 0.025	0.916 ± 0.027	0.996 ± 0.002	0.976 ± 0.005	0.976 ± 0.005	0.999 ± 0.000
	PhikonV2	0.947 ± 0.007	0.946 ± 0.008	0.996 ± 0.002	0.982 ± 0.004	0.982 ± 0.004	**1.000 ± 0.000**
	Dinobloom	0.938 ± 0.025	0.937 ± 0.026	0.989 ± 0.007	**0.991 ± 0.001**	**0.991 ± 0.001**	**1.000 ± 0.000**

for D3 where Accuracy increased from 0.519 to 0.746 and F1-score from 0.342 to 0.744); ResNet50 emerges as the most consistent performer across datasets with moderate improvements from fine-tuning, while ViT occupies an intermediate position with better generalization than MobileNetV3 but less consistency than ResNet50.

Experimental results on DL models revealed a significant finding: FMs may not be necessary for this particular task. Well-tuned, lightweight DL models can achieve performance comparable to larger FMs, offering a more practical and efficient option for clinical deployment in hematologic image classification (Table 2).

4 Conclusion

This study provides a thorough evaluation of four foundation models: GigaPath, PLIP, PhikonV2, and Dinobloom, for hematopathology classification across multiple datasets and experimental setups. The findings highlight key performance trends, strengths, and limitations of each model within the specialized context of digital pathology. Among them, Dinobloom consistently delivered superior results, with several configurations achieving accuracy, F1-score, and AUC values above 0.95. Its strong performance is likely attributed to an architecture well-suited for cellular-level image analysis. The observed variability across models and configurations highlights the importance of thoughtful model selection, dataset curation, and configuration tuning in clinical applications. Future work should explore the versatility of these models across a broader range of clinically relevant tasks and incorporate a wider array of deep learning-based models, particularly those specifically designed for medical image classification, to fully assess their utility in real-world medical imaging workflows. Overall, foundation models show significant promise for advancing hematopathology diagnostics, offering potential solutions to workforce shortages and rising diagnostic complexity when appropriately validated and deployed.

Disclosure of Interests. The authors declare no conflicts of interest.

References

1. ALL-IDB Acute Lymphoblastic Leukemia Image Database for Image Processing—scotti.di.unimi.it. https://scotti.di.unimi.it/all/. Accessed 26 May 2025
2. Acevedo, A.: A dataset for microscopic peripheral blood cell images for development of automatic recognition systems (2020)
3. Aria, M., Ghaderzadeh, M., Bashash, D., Abolghasemi, H., Asadi, F., Hosseini, A.: Acute lymphoblastic leukemia (ALL) image dataset (2021)
4. Ben Rabah, C., Petropoulos, I.N., Malik, R.A., Serag, A.: Vision transformers for automated detection of diabetic peripheral neuropathy in corneal confocal microscopy images. Front. Imaging **4**, 1542128 (2025)
5. Bian, Y., Li, J., Ye, C., Jia, X., Yang, Q.: Artificial intelligence in medical imaging: from task-specific models to large-scale foundation models. Chin. Med. J. **138**(06), 651–663 (2025)
6. Bodzas, A., Kodytek, P., Žídek, J.: A high-resolution large-scale dataset of pathological and normal white blood cells (2023)
7. Dehkharghanian, T., Mu, Y., Tizhoosh, H.R., Campbell, C.J.: Applied machine learning in hematopathology. Int. J. Lab. Hematol. **45**, 87–94 (2023)
8. Dosovitskiy, A., et al.: An image is worth 16×16 words: transformers for image recognition at scale. arXiv preprint arXiv:2010.11929 (2020)

9. El Achi, H., Khoury, J.D.: Artificial intelligence and digital microscopy applications in diagnostic hematopathology. Cancers **12**(4), 797 (2020)
10. Filiot, A., Jacob, P., Mac Kain, A., Saillard, C.: Phikon-v2, a large and public feature extractor for biomarker prediction. arXiv preprint arXiv:2409.09173 (2024)
11. Hacking, S.: Foundation models in pathology: bridging AI innovation and clinical practice (2025)
12. He, K., Zhang, X., Ren, S., Sun, J.: Deep residual learning for image recognition. In: Proceedings of the IEEE Conference on Computer Vision and Pattern Recognition, pp. 770–778 (2016)
13. Huang, Z., Bianchi, F., Yuksekgonul, M., Montine, T.J., Zou, J.: A visual-language foundation model for pathology image analysis using medical twitter. Nat. Med. **29**(9), 2307–2316 (2023)
14. Koch, V., et al.: DinoBloom: a foundation model for generalizable cell embeddings in hematology. In: Linguraru, M.G., et al. (eds.) MICCAI 2024. LNCS, vol. 15012, pp. 520–530. Springer, Cham (2024). https://doi.org/10.1007/978-3-031-72390-2_49
15. Koonce, B.: MobileNetV3. In: Koonce, B. (ed.) Convolutional Neural Networks with Swift for Tensorflow, pp. 125–144. Apress, Berkeley, CA (2021). https://doi.org/10.1007/978-1-4842-6168-2_11
16. Manzari, O.N., Ahmadabadi, H., Kashiani, H., Shokouhi, S.B., Ayatollahi, A.: MedViT: a robust vision transformer for generalized medical image classification. Comput. Biol. Med. **157**, 106791 (2023)
17. Metter, D.M., Colgan, T.J., Leung, S.T., Timmons, C.F., Park, J.Y.: Trends in the us and Canadian pathologist workforces from 2007 to 2017. JAMA Netw. Open **2**(5), e194337–e194337 (2019)
18. Russakovsky, O., et al.: ImageNet large scale visual recognition challenge. Int. J. Comput. Vision **115**, 211–252 (2015)
19. Serag, A., et al.: Translational AI and deep learning in diagnostic pathology. Front. Med. **6**, 185 (2019)
20. Walsh, E., Orsi, N.M.: The current troubled state of the global pathology workforce: a concise review. Diagn. Pathol. **19**(1), 163 (2024)
21. Wattenberg, M., Viégas, F., Johnson, I.: How to use t-SNE effectively. Distill **1**(10), e2 (2016)
22. Xu, H., et al.: A whole-slide foundation model for digital pathology from real-world data. Nature **630**(8015), 181–188 (2024)
23. Zheng, X., Wang, Y., Wang, G., Liu, J.: Fast and robust segmentation of white blood cell images by self-supervised learning. Micron **107**, 55–71 (2018). https://doi.org/10.1016/j.micron.2018.01.010, https://www.sciencedirect.com/science/article/pii/S0968432817303037

EndoTracker: Robustly Tracking Any Point in Endoscopic Surgical Scene

Lalithkumar Seenivasan[1](✉), Joanna Cheng[1](✉), Jin Fang[1,2](✉), Jin Bai[1], Roger D. Soberanis-Mukul[1], Jan Emily Mangulabnan[1], S. Swaroop Vedula[1], Masaru Ishii[3], Gregory Hager[1], Russell H. Taylor[1], and Mathias Unberath[1]

[1] Johns Hopkins University, Baltimore, MD, USA
{lseeniv1,unberath}@jhu.edu
[2] Hefei University of Technology, Hefei, Anhui, China
[3] Johns Hopkins Medical Institutions, Baltimore, MD, USA

Abstract. The most exciting frontiers of contemporary endoscopic video processing include tissue and instrument tracking and scene reconstruction together with their applications in surgical navigation, mixed reality visualization, and surgical automation. These tasks are enabled by task-specific techniques such as segmentation, object tracking, structure from motion, simultaneous localization and mapping (SfM and SLAM), and neural rendering. While these techniques vary in purpose and methodology, fundamentally, they rely on the ability to reliably and robustly establish point correspondences across video frames. However, point tracking in endoscopic scenes is hard due to the lack of distinct yet repetitive features, varying illuminations, and continuously changing visual appearance of corresponding points. A dense point-tracking model capable of reliably establishing point correspondences across video frames could catalyze endoscopic video processing and its downstream applications, and drive significant advancement in surgical data science. While any point-tracking foundation models have recently been proposed, they are trained on large simulated and natural scene videos and need to be adapted to endoscopic scenes to address domain-specific challenges. In this work, we present a dense point tracking foundation model for endoscopic scenes by fine-tuning a public foundation model on a large custom dataset comprising $13k$ endoscopic video sequences ($314k$ frames). We present three benchmark datasets with ground-truth point correspondences to quantitatively evaluate the point tracking performance in variable endoscopic scenes. Through quantitative analysis, we find that models fine-tuned on endoscopic scenes outperform out-of-the-box models, especially when considering conservative thresholds for tracking success, suggesting improved suitability for downstream tasks due to error propagation.

Keywords: Tracking Any Point · Foundation Model · Deep Learning

1 Introduction

Tissue and instrument tracking, and scene reconstruction are at the forefront of endoscopic video processing, spearheading the transformation in surgical navigation [21], mixed reality visualization [16], and surgical automation [3]. These tasks are currently enabled by contemporary task-specific techniques that include segmentation [4,19], object detection [9], tool-tip detection [17], structure from motion (SfM), and simultaneous localization and mapping [20]. Despite their varied objectives and approaches, these techniques fundamentally rely on accurately and consistently establishing temporal point correspondences across video frames. Methods addressing these fundamental needs – any point-tracking – could potentially revolutionize the approaches in enabling a wide range of downstream tasks in the surgical domain. The introduction of any point tracking foundation models in the computer vision domain, such as PIPs [11], TAPIR [8], and CoTracker [13,14], has initiated its rapid adoption with the surgical domain, with downstream tasks focusing on tracking instrument and tissues [22], and surgical navigation [20]. While these point tracking models show great promise as a key enabler for surgical applications, their adoption for clinical applications is severely hindered by domain shifts and domain-specific challenges.

The introduction of the TAPVid large-scale dataset [7] and Kubric synthetic dataset [10] has catalyzed the advancement of any point tracking foundation models [8,13,14]. Trained on massive synthetic datasets, these models allow tracking of any points in natural dynamic scenes. Although they exhibit generalizability to the surgical domain to an extent, with recent works adopting these models to aid in instrument tracking [22] and surgical navigation [20], they are severely limited by domain shift and domain-specific challenges. Firstly, unlike natural scenes, tissue surfaces exhibit homogeneous visual characteristics, often lacking distinct features for the model to reason and establish accurate point correspondences across frames. Secondly, surgical scenes suffer from temporal photometric inconsistency due to varying lighting conditions, causing the same tissue surfaces to exhibit different textures and reflections across frames.

To this end, we first custom-curate a large endoscopic dataset - comprising of 13k short endoscopic video sequences (each consisting of 24 frames) - to fine-tune any-point tracking foundation models in a self-supervised setting. To quantitatively evaluate point-tracking models on endoscopic scenes, we additionally introduce 3 point-tracking benchmarking datasets. We further present EndoTracker–an adaptation of CoTracker3 [13] fine-tuned on our meticulously curated large endoscopic dataset—designed to mitigate domain shifts and serve as a catalyst for advancing endoscopic video processing and its application. Through quantitative analysis, we show that the EndoTracker performs significantly better than the out-of-the-box CoTracker3 foundation model (trained and fine-tuned on computer-vision datasets).

2 Data Curation

Large Endoscopic Dataset for Any Point Tracking: Our dataset for fine-tuning the foundation model was curated from five endoscopic datasets, carefully chosen to potentially help bridge domain challenges in endoscopic surgery and navigation. These include (i) an in-house sinus endoscopic navigation dataset generated from 7 cadaveric studies, and four public datasets: (ii) SimCol3D [18] – a synthetic colon dataset, (iii) StereoMIS [12] – dataset for SLAM in endoscopic videos, (iv) Bronchoscopy [6] – benchmarking dataset for bronchoscopy odometry, and (v) PitVis2023 [5] – workflow recognition dataset for pitutery surgery. Given the domain challenges in endoscopic navigation and surgical scenes - such as dynamic scene changes as the endoscope continues to traverse, and partial/full occlusions from instruments/irrigation/blood- that could potentially hinder domain-adapting the point tracking model in a self-supervised setting, the videos from all five datasets were first segmented into short video sequences of 24 frames each. All short sequences were then meticulously filtered by manual review to ensure that at least a partial tissue region remained visible across all frames, ensuring the presence of temporal point correspondences. In total, 13,170 video sequences (comprising > 314k frames) were curated. Figure 1(a) shows the distribution and samples of the video sequences from the five datasets.

Endoscopic Benchmarking Datasets: To enable quantitative evaluation of the model's performance in tracking points in endoscopic scenes, we curate point tracks for three publicly available datasets. **(i) C3VD:** A test set comprising 409 video sequences with ground truth point tracks is generated from a public synthetic colonoscopy 3D video dataset [1]. To generate ground truth point tracks

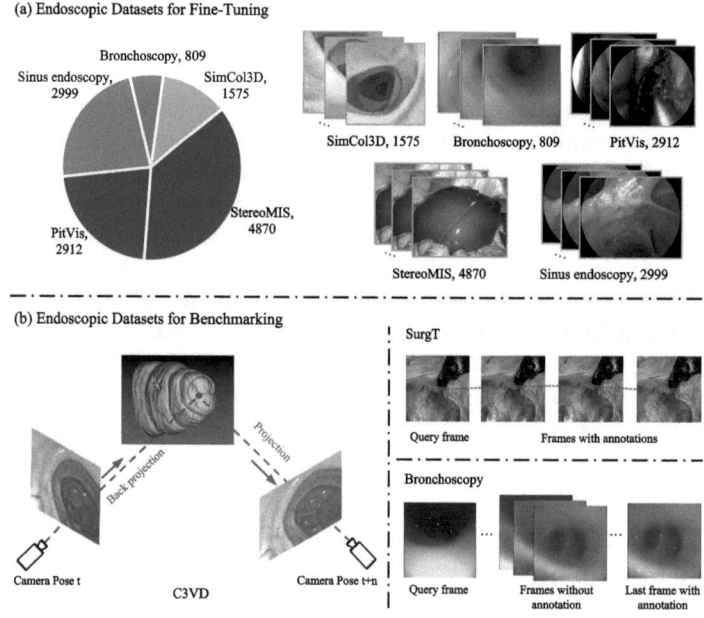

Fig. 1. Dataset overview: (a) Distribution and visualization of our custom-curated endoscopic dataset showing diversity across surgical procedures and tissue types. (b) Construction methodology of our benchmarks (C3VD, SurgT, and Bronchoscopy) utilizing 3D anatomical models and precise camera pose estimation to generate ground truth correspondences for quantitative evaluation.

across frames, we utilize the 3D model and camera poses from the dataset. We first generate depth maps for camera frames (RGB image) using the 3D model and camera pose. Then, using the depth map and camera pose, a pixel point from one camera frame is first back-projected to a point in the 3D space and then forward-projected to the next camera frame, establishing point correspondences between frames. This process is then iterated for multiple points across 24 frames to generate ground truth point tracks for a video sample. **(ii) SurgT:** A surgical tracking public dataset with bounding box annotation [2]. We use its 125 videos to curate 1768 video sequences as the second held-out test set. To generate ground truth point tracks, we utilize the bounding box annotation, using the bounding box's center as the ground truth point track across all frames. **(iii) Bronchoscopy:** As bronchoscope data is one of the challenging data to establish point correspondence, we set aside 12 video sequences generated from the Bronchoscopy dataset [6] as the third held-out test set. For this test set, we manually annotated 5 points per sequence, for the first and last frame. Figure 1(b) illustrates the construction methodology of endoscopic benchmarking datasets.

3 EndoTracker

3.1 Preliminary

CoTracker3 [13] is a transformer-based dense point tracking foundation model. Given query points $P_i^{t_i} = (x_i^{t_i}, y_i^{t_i}) \in \mathbb{R}^2$ at frames t_i for $i \in \{1, 2, \ldots, N\}$, the model estimates their positions across all frames. The model is initially trained on synthetic data in a fully supervised manner and then fine-tuned on natural videos in a self-supervised setting employing a teacher-student framework. During inference, the model computes convolutional features per frame and 4D correlations between query point features and all frames. A transformer processes a token grid where for each query point $i = 1, \ldots, N$ and time step t, it concatenate bidirectional track embeddings $(\eta_{t-1 \to t}, \eta_{t \to t+1})$, confidence C_i^t, visibility V_i^t, and 4D correlations Corr_i^t. Using factorized time attention, group attention across points, and proxy tokens for efficiency, the transformer iteratively refines the tracks, confidence, and visibility through M update steps: $P^{(m+1)} = P^{(m)} + \Delta P^{(m+1)}$, $C^{(m+1)} = C^{(m)} + \Delta C^{(m+1)}$, and $V^{(m+1)} = V^{(m)} + \Delta V^{(m+1)}$. The model offers two modes: an online version for forward-only tracking and an offline version for bidirectional tracking.

3.2 Fine-Tuning on Endoscopic Scenes

To domain adapt CoTracker3 to endoscopic scenes, we adopt the original CoTracker3 training regime. In the initial fully-supervised training phase, we leverage the Kubric dataset [10]—a public dataset with comprehensive ground truth point track annotations. Then, extending the model to the endoscopic domain, we fine-tuned the model on our large custom-curated endoscopic dataset in a self-supervised manner using the teacher-student framework. We use TAPIR [8], CoTracker3-online, CoTracker3-offline trained with window size 12, and CoTracker2 [14] as teacher models. However, unlike the original self-supervised training protocol, where points are first initialized using SIFT descriptors [15], we adopt a uniform grid-based initialization strategy. This choice is motivated by the domain-specific challenges. As the endoscopic scenes often lack distant features on tissue surfaces, the number of points that can be initialized using feature-based methods, such as SIFT descriptors [15], is limited. Therefore, to ensure sufficient points are initialized in the self-supervised training regime, we adopted a uniform grid-based initialization strategy.

4 Experimental Setup

4.1 Datasets

Fully-Supervised Pre-training Dataset: Following the original implementation of CoTracker3, EndoTracker is pre-trained on 6k video samples with ground truth point tracks generated from the Kurbic dataset [10]. **Self-supervised fine-tuning dataset:** The pre-trained model is then finetuned on our custom-curated endoscopic datasets (13k video sequences), with pseudo-annotations generated using SOTA point tracking models in a teacher-student framework. These 13k video sequences include all 2999 video sequences generated from the in-house sinus endoscopy dataset, all 1575 video sequences from the SimCol3D, 2912 video sequences from PitVis, 4821 video sequences from StereoMIS, and 795 sequences from the Bronchoscopy dataset. **Test set:** The test set for quantitative analysis includes 409 video sequences generated from C3VD, 1768 sequences from SurgT, and 14 held-out test sequences from the Bronchoscopy dataset. Additionally, 10 video sequences from PitVis and 49 sequences from StereoMIS are used as the held-out test set for qualitative analysis.

4.2 Implementation Details

In this work, we primarily focus on the online mode of the CoTracker3. All models were trained using 8 NVIDIA H100 GPUs on the Johns Hopkins Discovery HPC cluster. The training process consisted of two phases: (1) fully-supervised pre-training on ∼6k Kubric videos for 200k steps, followed by (2) self-supervised fine-tuning on our ∼13k custom-curated endoscopic dataset for 60k steps. Across all fully-supervised pre-training regime, we consistently utilized 384 point tracks per video. During fine-tuning, we implemented uniform sampling for query point initialization, maintaining the same density of 384 query points per video sequence. All fine-tuning was conducted with a learning rate of 5×10^{-4}.

5 Results and Discussion

Focusing on the online model, the EndoTracker is benchmarked against out-of-the-box CoTracker2 [14] and CoTracker3 [13] quantitatively on all three of

Table 1. Comparison of EndoTracker against state-of-the-art models in online mode (out of the box), with * denoting statistical significance ($p < 0.05$) of EndoTracker's performance over CoTracker3. OA: Occlusion Accuracy; C: Confidence.

Model	OA	Jaccard (J)						Points within x pixels(δ_x)						C
		J_1	J_2	J_4	J_8	J_{16}	J_{avg}	δ_1	δ_2	δ_4	δ_8	δ_{16}	δ_{avg}	
C3VD														
CoTracker2	98.41	69.03	88.20	96.36	98.21	98.39	90.04	78.91	93.22	98.61	99.80	99.95	94.10	/
CoTracker3	99.69	68.41	88.89	97.37	99.38	99.65	90.74	78.41	93.26	98.62	99.77	99.94	94.00	99.93
EndoTracker	99.53	70.21*	89.64*	97.53*	99.29	99.50	91.23*	79.52*	93.70*	98.79*	99.83*	99.96*	94.36*	99.90*
SurgT														
CoTracker2	98.31	87.84	97.54	98.10	98.22	98.22	95.99	92.59	99.55	99.94	100.00	100.00	98.41	/
CoTracker3	98.51	96.93	98.32	98.32	98.32	98.32	98.04	99.04	100.00	100.00	100.00	100.00	99.81	99.57
EndoTracker	98.38	96.99	98.30	98.30	98.30	98.30	98.04	99.09	100.00	100.00	100.00	100.00	99.82	99.37*
Bronchoscopy														
CoTracker2	78.59	2.20	2.20	5.48	18.26	35.60	12.75	3.46	3.46	9.62	26.15	45.00	17.54	/
CoTracker3	69.10	2.06	4.62	8.13	22.64	44.62	16.41	3.46	7.31	12.31	28.08	65.51	23.33	80.39
EndoTracker	67.56	5.86	8.24	12.46	24.78	45.77	19.42	8.85	14.23	19.62	40.13	63.21	29.21	84.65*

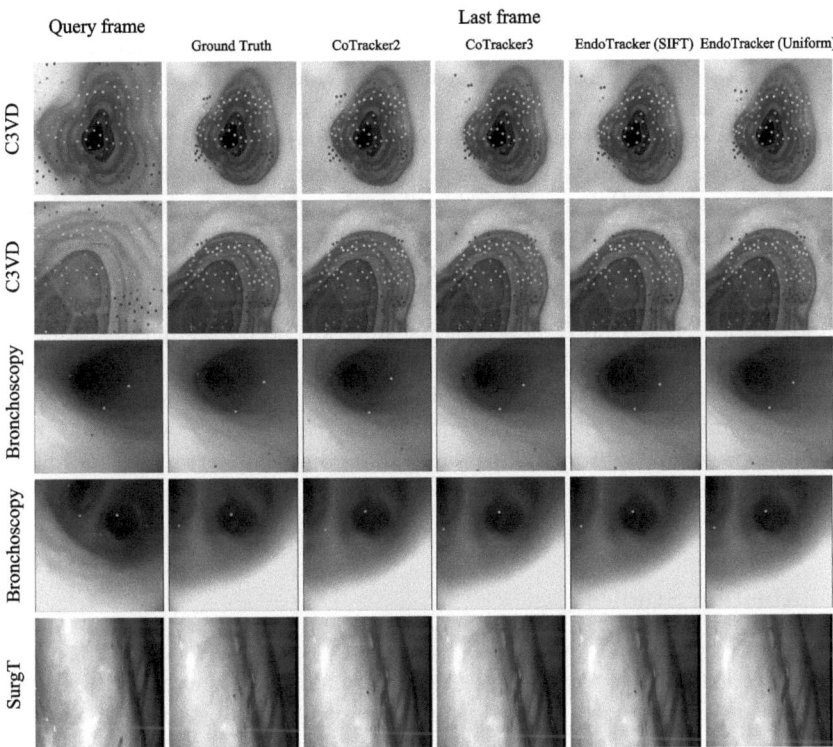

Fig. 2. Qualitative comparison of tracking results on three benchmarks (C3VD, Bronchoscopy, SurgT). The leftmost columns show the initial query points on the reference frame, while subsequent columns show the estimated point positions on the final frame of the sequence.

our test sets (Table 1). While the Jaccard similarity (J) and points within x pixels (δ_x) were calculated across all frames for video sequences curated from C3VD and SurgT, we only calculated their scores based on the last manually annotated frame for the video sequences curated from the Broncoscopy dataset. Based on the paired t-test, we find that EndoTracker significantly (with statistical significance $p < 0.05$) outperforms out-of-the-box CoTracker3 on the held-out C3VD dataset. Specifically, significant improvement was observed in conservative Jaccard similarity (J_1, J_2, and J_4) and points within x pixels (δ_1, δ_2, and δ_4). Achieving better performance in these conservative matrics is crucial for enhancing surgical navigation across challenging anatomical structures. However, no statistically significant improvement in performance is observed in the SurgT or Bronchoscopy dataset. Computationally, the EndoTracker model takes on average ~ 0.08 s to inference point-tracks for each window on bronchoscopy dataset, faclitating online inference at ~ 12.5 frames per second.

The lack of significant performance improvement in SurgT dataset could be partly attributed to the presence of clear and distinct visual features present across its scenes, resulting in the out-of-the-box CoTracker3 model achieving very high performance. Additionally, each of its video sequences only has 1 track, reducing the sample size. The general low performance of all models in the bronchoscopy data could be potentially attributed to the extreme scarcity of distinctive visual features on tissue surfaces, which, in our view, necessitates targeted updates to the model architecture in addition to incorporating our domain-bridging datasets.

Figure 2 shows qualitative comparisons across all evaluated models. On the C3VD dataset, CoTracker2 demonstrates the best overall alignment with ground truth, while EndoTracker shows superior point predictions compared to CoTracker3. In the Bronchoscopy dataset, characterized by fewer and more sparsely distributed tracking points, EndoTracker (uniform sampling) clearly exhibits enhanced position accuracy despite the challenging low-texture, homogeneous appearance of bronchial passages that complicates tracking for all models. For the SurgT dataset, all models except CoTracker3 display comparable performance.

5.1 Ablation Studies

We conducted two ablation studies to evaluate EndoTracker. First, we evaluate query point initialization strategies for tracking performance (Table 2). We observed that EndoTracker trained using a uniform sampling strategy performed significantly better on the C3VD dataset than when trained using a SIFT-based sampling strategy. This could potentially be a result of EndoTracker being fine-tuned on lesser point tracks in the SIFT-based sampling strategy as SIFT may fail to generate a sufficient number of key points in endoscopic scenes due to a lack of distinct features. Uniform sampling forces the EndoTracker to train on a fixed number of samples.

Second, we evaluated the efficacy of domain-specific fine-tuning on our endoscopic dataset (Table 3). EndoTracker consistently outperforms its pre-trained

Table 2. Ablation study on query point initialization strategies: Comparison between uniform sampling and SIFT-based keypoint extraction. The symbol * denotes statistical significance ($p < 0.05$) of EndoTracker's performance when trained using uniform sampling over SIFT-based sampling strategy.

Sampling Strategy	OA	Jaccard (J)						Points within x pixels (δ_x)						C
		J_1	J_2	J_4	J_8	J_{16}	J_{avg}	δ_1	δ_2	δ_4	δ_8	δ_{16}	δ_{avg}	
C3VD														
SIFT	99.34	68.91	88.94	97.16	99.04	99.30	90.67	78.79	93.45	98.70	99.79	99.95	94.14	99.80
Uniform	99.53*	70.21*	89.64*	97.53*	99.29*	99.50*	91.23*	79.52*	93.70*	98.79*	99.83*	99.96*	94.36*	99.90*
SurgT														
SIFT	98.77	97.34	98.51	98.51	98.51	98.51	98.27	99.21	100.00	100.00	100.00	100.00	99.84	99.24
Uniform	98.38	96.99*	98.30	98.30	98.30	98.30	98.04	99.09*	100.00	100.00	100.00	100.00	99.82*	99.37
Bronchoscopy														
SIFT	66.41	2.20	5.43	12.28	27.92	43.24	18.22	3.46	8.46	22.18	43.97	62.44	28.10	81.68
Uniform	67.56	5.86	8.24	12.46	24.78	45.77	19.42	8.85	14.23	19.62	40.13	63.21	29.21	84.65

Table 3. Ablation study on fine-tuning: Performance comparison between our EndoTracker and the original pre-trained CoTracker3. The symbol * denotes statistical significance ($p < 0.05$) of EndoTracker from fine-tuning on endoscopic video sequences.

Sampling Strategy	OA	Jaccard (J)						Points within x pixels (δ_x)						C
		J_1	J_2	J_4	J_8	J_{16}	J_{avg}	δ_1	δ_2	δ_4	δ_8	δ_{16}	δ_{avg}	
C3VD														
CoTracker3 w/o fine-tuning	99.08	57.85	82.5	94.43	98.08	98.87	86.35	70.32	89.15	96.83	99.07	99.7	91.01	99.32
EndoTracker	99.53*	70.21*	89.64*	97.53*	99.29*	99.50*	91.23*	79.52*	93.70*	98.79*	99.83*	99.96*	94.36*	99.90*
SurgT														
CoTracker3 w/o fine-tuning	98.83	96.78	97.58	97.63	97.63	98.06	97.54	98.69	99.29	99.34	99.34	99.69	99.27	99.67
EndoTracker	98.38*	96.99*	98.30*	98.30*	98.30*	98.30*	98.04*	99.09*	100.00*	100.00*	100.00*	100.00*	99.82*	99.37
Bronchoscopy														
CoTracker3 w/o fine-tuning	66.92	0.85	1.95	1.95	3.88	9.57	3.64	1.54	3.46	3.46	6.92	13.85	5.85	86.26
EndoTracker	67.56	5.86	8.24*	12.46*	24.78*	45.77*	19.42*	8.85	14.23*	19.62*	40.13*	63.21*	29.21*	84.65

counterparts across all metrics. On the C3VD and Bronchoscopy datasets which contain more challenging scenarios, our EndoTracker exhibits notable improvements. This performance gain highlights the quality of our custom-curated funetuning dataset and its ability to serve as a bridge between the computer vision and endoscopic domains.

6 Conclusion

Track-Any-Point has the potential to play a catalytic role in transforming endoscopic navigation, surgical automation, and intraoperative decision-making. However, the direct adoption of out-of-the-box computer vision models is potentially hindered by domain-specific challenges in endoscopic scenes. Introducing carefully curated point tracking benchmarking and fine-tuning datasets for endoscopic scenes, this work potentially serves as a bridge in adapting the transformative impact of any point tracking model from computer vision to the endoscopic domain. We also introduce EndoTracker, an adaptation of CoTracker3 for endoscopic scenes. Through extensive qualitative and quantitative analysis, we demonstrate that fine-tuning models on our endoscopic dataset effectively addresses most of the domain shift problem that occurs when models trained on natural images are applied to endoscopic sequences. Nonetheless, challenges persist in extreme cases—such as bronchoscopy data—where all state-of-the-art models struggle due to the severe lack of distinctive visual features. Our custom-curated endoscopic dataset and EndoTracker serve as a valuable foundation for advancing endoscopic video processing and enabling more robust applications in surgical settings.

Acknowledgement. This work was funded in part by Johns Hopkins University internal funds and in part by NIH R01EB030511. The content is solely the responsibility of the authors and does not necessarily represent the official views of the National Institutes of Health.

References

1. Bobrow, T.L., Golhar, M., Vijayan, R., Akshintala, V.S., Garcia, J.R., Durr, N.J.: Colonoscopy 3D video dataset with paired depth from 2D-3D registration. Med. Image Anal. 102956 (2023)
2. Cartucho, J., et al.: SurgT challenge: benchmark of soft-tissue trackers for robotic surgery. Med. Image Anal. **91**, 102985 (2024)
3. Chadebecq, F., Lovat, L.B., Stoyanov, D.: Artificial intelligence and automation in endoscopy and surgery. Nat. Rev. Gastroenterol. Hepatol. **20**(3), 171–182 (2023)
4. Colleoni, E., Edwards, P., Stoyanov, D.: Synthetic and real inputs for tool segmentation in robotic surgery. In: Martel, A.L., et al. (eds.) MICCAI 2020. LNCS, vol. 12263, pp. 700–710. Springer, Cham (2020). https://doi.org/10.1007/978-3-030-59716-0_67
5. Das, A., et al.: PitVis-2023 challenge: workflow recognition in videos of endoscopic pituitary surgery. arXiv preprint arXiv:2409.01184 (2024)
6. Deng, J., Li, P., Dhaliwal, K., Lu, C.X., Khadem, M.: Feature-based visual odometry for bronchoscopy: a dataset and benchmark. In: 2023 IEEE/RSJ International Conference on Intelligent Robots and Systems (IROS), pp. 6557–6564. IEEE (2023)
7. Doersch, C., et al.: TAP-vid: a benchmark for tracking any point in a video. Adv. Neural. Inf. Process. Syst. **35**, 13610–13626 (2022)
8. Doersch, C., et al.: TAPIR: tracking any point with per-frame initialization and temporal refinement. In: Proceedings of the IEEE/CVF International Conference on Computer Vision, pp. 10061–10072 (2023)
9. Fujii, R., Hachiuma, R., Kajita, H., Saito, H.: Surgical tool detection in open surgery videos. Appl. Sci. **12**(20), 10473 (2022)
10. Greff, K., et al.: Kubric: a scalable dataset generator (2022)
11. Harley, A.W., Fang, Z., Fragkiadaki, K.: Particle video revisited: tracking through occlusions using point trajectories. In: ECCV (2022)
12. Hayoz, M., et al.: Learning how to robustly estimate camera pose in endoscopic videos. Int. J. Comput. Assist. Radiol. Surg. 1–8 (2023). https://doi.org/10.1007/s11548-023-02919-w
13. Karaev, N., Makarov, I., Wang, J., Neverova, N., Vedaldi, A., Rupprecht, C.: Cotracker3: simpler and better point tracking by pseudo-labelling real videos (2024). https://arxiv.org/abs/2410.11831

14. Karaev, N., Rocco, I., Graham, B., Neverova, N., Vedaldi, A., Rupprecht, C.: CoTracker: it is better to track together (2024). https://arxiv.org/abs/2307.07635
15. Lowe, D.G.: Distinctive image features from scale-invariant keypoints. Int. J. Comput. Vision **60**, 91–110 (2004)
16. Morimoto, T., et al.: XR (extended reality: virtual reality, augmented reality, mixed reality) technology in spine medicine: status quo and quo vadis. J. Clin. Med. **11**(2), 470 (2022)
17. Park, J., Park, C.H.: Recognition and prediction of surgical actions based on online robotic tool detection. IEEE Robot. Autom. Lett. **6**(2), 2365–2372 (2021)
18. Rau, A., et al.: SimCol3D–3D reconstruction during colonoscopy challenge. Med. Image Anal. **96**, 103195 (2024)
19. Ravi, N., et al.: SAM 2: segment anything in images and videos. arXiv preprint arXiv:2408.00714 (2024)
20. Teufel, T., et al.: OneSLAM to map them all: a generalized approach to slam for monocular endoscopic imaging based on tracking any point. Int. J. Comput. Assist. Radiol. Surg. 1–8 (2024)
21. Tzelnick, S., et al.: Skull-base surgery-a narrative review on current approaches and future developments in surgical navigation. J. Clin. Med. **12**(7), 2706 (2023)
22. Zhan, B., et al.: Tracking everything in robotic-assisted surgery. arXiv preprint arXiv:2409.19821 (2024)

Temporally-Constrained Video Reasoning Segmentation and Automated Benchmark Construction

Yiqing Shen[ID], Chenjia Li, Chenxiao Fan, and Mathias Unberath[✉]

Johns Hopkins University, Baltimore, MD, USA
{yshen92,unberath}@jhu.edu

Abstract. Conventional approaches to video segmentation are confined to predefined object categories and cannot identify out-of-vocabulary objects, let alone objects that are not identified explicitly but only referred to implicitly in complex text queries. This shortcoming limits the utility for video segmentation in complex and variable scenarios, where a closed set of object categories is difficult to define and where users may not know the exact object category that will appear in the video. Such scenarios can arise in operating room video analysis, where different health systems may use different workflows and instrumentation, requiring flexible solutions for video analysis. Reasoning segmentation (RS) now offers promise towards such a solution, enabling natural language text queries as interaction for identifying object to segment. However, existing video RS formulation assume that target objects remain contextually relevant throughout entire video sequences. This assumption is inadequate for real-world scenarios in which objects of interest appear, disappear or change relevance dynamically based on temporal context, such as surgical instruments that become relevant only during specific procedural phases or anatomical structures that gain importance at particular moments during surgery. To enable more research on RS for dynamic tasks, our first contribution is the introduction of **temporally-constrained video reasoning segmentation**, a novel task formulation that requires models to implicitly infer when target objects become contextually relevant based on text queries that incorporate temporal reasoning. However, we do not know how well method perform this task, because we do not have a dataset to study this. So the first step is to construct a dataset. Since manual annotation of temporally-constrained video RS datasets would be expensive and limit scalability, our second contribution is an innovative automated benchmark construction method. Finally, we present *TCVideoRSBenchmark*, a temporally-constrained video RS dataset containing 52 samples using the videos from the MVOR dataset. The TCVideoRSBenchmark is available at https://github.com/arcadelab/TCVideoRSBenchmark.

Keywords: Video Analysis · Reasoning Segmentation · Digital Twin Representation · Large Language Model (LLM) Agent · Benchmark

1 Introduction

Conventional video segmentation task formulations, including semantic segmentation and instance segmentation, are fundamentally limited by their confinement to predefined object categories and their inability to respond to text queries that require understanding of implicit relationships and multi-step reasoning for object identification [3]. These limitations restrict their applicability in dynamic clinical environments, such as operating room (OR) video analysis for monitoring surgical workflow, which requires the ability to respond to context-dependent queries that go beyond simple object identification, encompassing complex procedural understanding that traditional segmentation methods cannot provide. Reasoning segmentation (RS) [5] enables text-based object identification and has shown promise to enhance user interaction in surgical workflow analysis [10,11]. However, existing video RS methods operate under a critical assumption that target objects remain contextually relevant throughout entire video sequences. This assumption becomes inadequate for real-world applications where objects of interest appear, disappear, or change relevance dynamically based on temporal context. In other words, current video RS approaches cannot effectively handle queries such as *"segment the anesthesia equipment only during the patient preparation phase"* that require understanding of temporal boundaries [11].

This temporal limitation undermines the potential of RS for applications that require precise temporal understanding, where these video monitoring frameworks must understand not only what and where objects are located, but also precisely when they become relevant within specific procedural contexts [4,13]. In surgical workflows, for instance, procedures exhibit inherently structured temporal organization with distinct phases such as patient preparation, anesthesia induction, surgical intervention, and recovery, each characterized by different sets of relevant objects and personnel configurations [2]. The importance of temporal relationships extends beyond simple phase identification to encompass complex dependencies between procedural events and object relevance periods, where the same instrument or personnel may require different analytical attention depending on the current procedural context. Despite this need, the temporal dimension remains largely unexplored in current RS literature due to the absence of appropriate benchmarks.

To address this gap, we first introduce a novel task formulation termed **temporally-constrained video reasoning segmentation**, as illustrated in Fig. 1. This task formulation extends video RS beyond continuous object tracking by incorporating phase-specific or action-specific temporal constraints to perform segmentation. For this new task, due to the lack of appropriate dataset, we do not know how model performs. Consequently, the initial step is to construct a benchmark dataset. Correspondingly, we propose an automated benchmark construction method that leverages digital twin (DT) representations, defined as structured intermediate representations that preserve semantic, spatial, and temporal relationships between entities and their interactions [8], combined with large language models (LLMs) to generate temporally aware implicit queries without requiring manual annotation efforts that would otherwise limit the scalability of the dataset. Unlike previous applications of digital twin representations that primarily utilized semantic and spatial information for general reasoning tasks [9], our approach specifically exploits the temporal dimension embedded within DT structures to construct queries that require understanding of when objects become relevant within surgical workflow phases, enabling the generation of temporally-constrained reasoning queries that reflect the dynamic nature of OR procedures.

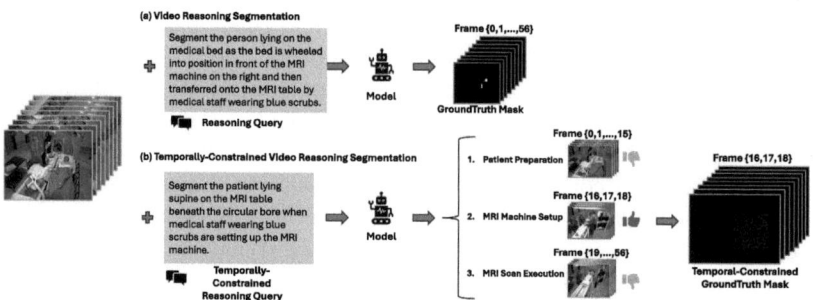

Fig. 1. Comparison between conventional video RS and the proposed temporally-constrained video RS task formulation. (a) Conventional video RS processes implicit text queries across entire video sequences, generating segmentation masks for all frames regardless of temporal relevance. (b) Temporally-constrained video reasoning segmentation restricts segmentation to specific temporal boundaries. The example demonstrates segmenting a patient only during the "MRI Machine Setup" phase (frames 16–18) rather than throughout the entire video sequence.

The major contributions are three-fold. First, we propose the temporally-constrained video RS, which is a new task that requires models to perform RS only within specified temporal boundaries. Second, we develop an automated pipeline that constructs benchmark datasets through DT representations and LLM-based query generation, enabling the scalable creation of temporally-constrained reasoning queries. Third, we construct a benchmark dataset for temporally-constrained video RS (namely *TCVideoRSBenchmark*), which contains 52 samples that span various surgical scenarios and temporal reasoning.

2 Task Formulation

We formalize temporally-constrained video RS as an extension of traditional video RS that incorporates temporal boundaries derived from surgical workflow phases. Unlike conventional video RS that assumes the presence and relevance of continuous objects throughout entire video sequences, our formulation acknowledges that surgical procedures exhibit a structured temporal organization where objects, personnel, and equipment become contextually relevant only during specific procedural phases. Specifically, given an input video sequence $\mathcal{V} = \{I^{(1)}, I^{(2)}, \ldots, I^{(T)}\}$ consisting of frames T and an implicit reasoning query Q that describes the target segmentation objective, traditional video reasoning segmentation seeks to produce a sequence of binary segmentation masks $\mathcal{M} = \{M^{(1)}, M^{(2)}, \ldots, M^{(T)}\}$ where each $M^{(t)} \in \{0,1\}^{H \times W}$ indicates the object in pixels at the timestep t. However, this formulation assumes that the reasoning query Q remains equally applicable across all temporal instances, which proves inadequate for surgical workflow analysis where contextual relevance varies across procedural phases.

In contrast, temporally-constrained video RS requires models to implicitly infer temporal boundaries from the reasoning query itself, determining when the segmentation objective becomes contextually applicable without explicit temporal annotations. Therefore, the model must parse the implicit query Q to extract both the target segmentation objective and the underlying temporal constraints embedded within the query semantics. Formally, we define a temporal constraint $\tau_Q : \mathbb{N} \rightarrow \{0,1\}$ that the model must learn to determine the validity of the reasoning query at each time step based solely on the query content and the video context:

$$\tau_Q(t) = \begin{cases} 1 & \text{if } f_{\text{inference}}(Q, \mathcal{V}, t) = \text{active} \\ 0 & \text{otherwise} \end{cases} \quad (1)$$

where $f_{\text{inference}}$ represents the temporal reasoning capacity of the model that analyzes the query Q, the video sequence $\mathcal{V}$, and the current time step t to determine whether segmentation should be performed. The temporally-constrained reasoning segmentation task then becomes:

$$\mathcal{M}_{\text{constrained}} = \{M^{(t)} \cdot \tau_Q(t) | t = 1, 2, \ldots, T\}. \quad (2)$$

This definition ensures that segmentation masks are produced only during periods that the model infers as temporally relevant from the query, with $M^{(t)} = \emptyset$ (empty mask) when $\tau_Q(t) = 0$.

3 Dataset Construction

Digital Twin Representation for Temporal Reasoning. Although prior work on DT representations [9] for automatic reasoning data generation focuses on general visual reasoning tasks, our approach introduces surgical workflow-aware temporal modeling that explicitly captures temporal information such as procedural stage transitions and temporal constraints. Our dataset construction begins with the transformation of the OR video sequences into structured DT representations that not only preserve semantic and spatial relationships but also encode temporal boundaries for analysis of surgical workflow. Formally, given an OR video sequence $\mathcal{V} = \{I^{(1)}, I^{(2)}, \ldots, I^{(T)}\}$, we construct a corresponding DT representation $\mathcal{J} = \{\mathcal{J}^{(1)}, \mathcal{J}^{(2)}, \ldots, \mathcal{J}^{(T)}\}$ where each frame-level representation $\mathcal{J}^{(t)}$ encodes multi-dimensional information through a suite of vision

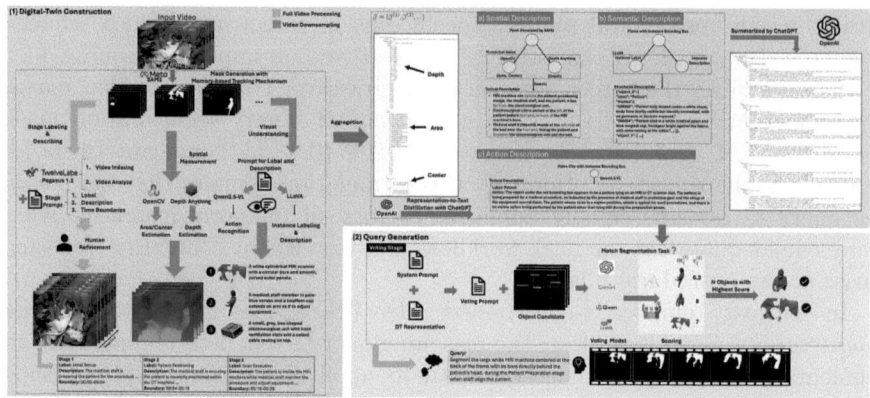

Fig. 2. Overview of the proposed automated pipeline for temporally-constrained video RS benchmark construction. The framework consists of two primary components: (1) Digital twin construction, which transforms raw operating room video sequences into structured representations through specialized vision foundation models, including TwelveLabs Pegasus 1.2 for action and phase identification, SAM [7] for instance segmentation, DepthAnything2 [14] for spatial measurement, and LLaVA [6] for semantic understanding; and (2) Query generation, which employs an ensemble voting by LLM to identify salient object candidates and their associated procedural phases from the digital twin representation, followed by template-based synthesis of temporally-constrained reasoning queries that embed implicit temporal boundaries within natural language formulations.

models $\Omega = \{\omega_1, \omega_2, \ldots, \omega_K\}$, which extract information, formally expressed as $\mathcal{J}^{(t)} = \Omega(I^{(t)})$. Temporal information extraction begins with the identification of the surgical stage at the video level through TwelveLabs Pegasus 1.2[1], a multimodal foundation model for video indexing and analysis to decompose input video sequences into distinct procedural phases $\Phi = \{\phi_1, \phi_2, \ldots, \phi_P\}$ in the full frame video stream. For each phase, TwelveLabs Pegasus 1.2 extracts three components, namely descriptive labels that categorize procedural activity, detailed descriptions that capture specific actions and interactions between objects that occur during the stage, and the corresponding temporal boundaries that define the temporal extent of each phase. Subsequently, this automated extraction is refined through human verification to ensure accuracy and clinical relevance. For semantic and spatial encoding, we employ SAM2 [7] for example identification and segmentation, generating masks $\mathcal{M}^{(t)} = \{m_i^{(t)}\}_{i=1}^{N^{(t)}}$ ($t = 1, \cdots, T$) where each $m_i^{(t)}$ represents a binary mask for object i with confidence score $\epsilon_i^{(t)}$ at time step t. The notation $N^{(t)}$ denotes the number of objects detected in frame t, and $\epsilon_i^{(t)} \in [0, 1]$ quantifies the confidence in the segmentation of object i derived from SAM2 [7]. For temporal coherence across frames, we leverage SAM2's memory-

---

[1] https://www.twelvelabs.io/blog/introducing-pegasus-1-2.

based tracking mechanism (Fig. 2):

$$m_i^{(t+k)} = \text{SAM}_{\text{track}}(I^{(t+k)}, \{m_i^{(t+k')}\}_{k'=0}^k), \quad 0 < k < t_s \quad (3)$$

where $I^{(t+k)}$ represents the target frame for tracking, $\{m_i^{(t+k')}\}_{k'=0}^k$ denotes the sequence of previous masks used for temporal propagation, and t_s represents the temporal sampling interval for key frame processing to balance computational efficiency with tracking accuracy. Additional spatial information is complemented $\mathcal{M}^{(t)}$ through DepthAnything2 [14], which generates dense depth maps $D^{(t)} \in \mathbb{R}^{H \times W}$ for each frame of resolution $H \times W$. For every object i, we can compute the depth statistics within its mask region as $d_i^{(t)} = \{D^{(t)}(p) | p \in m_i^{(t)}\}$, where p represents the coordinates of the pixels within the mask. The mean depth $\mu_i^{(t)} = \frac{1}{|m_i^{(t)}|} \sum_{p \in m_i^{(t)}} D^{(t)}(p)$ and the standard deviation $\sigma_i^{(t)} = \sqrt{\frac{1}{|m_i^{(t)}|} \sum_{p \in m_i^{(t)}} (D^{(t)}(p) - \mu_i^{(t)})^2}$ characterize the spatial positioning and the depth variation, where $|m_i^{(t)}|$ denotes the number of pixels in the mask. Finally, we also employ OpenCV operators to extract complementary visual features, including optical flow vectors for motion analysis, color histograms for appearance characterization, and texture descriptors following previous work [9]. We also include a description of temporal action at the object level to encode temporal information at the object level as a complement to semantic understanding and temporal information at the video level, where we generate stage-specific action sequences $\mathcal{A}^{(t)} = \{a_{i,j}^{(t)}\}_{j=1}^{M_i^{(t)}}$ for each object i via Qwen2.5-VL [1]. These action descriptions explicitly encode when specific activities begin, progress, and conclude, enabling the DT representation to understand temporal relevance of object within procedural contexts. Semantic understanding is achieved through LLaVA [6], which generates object-level descriptors $\mathcal{S}^{(t)} = \{s_i^{(t)}\}_{i=1}^{N^{(t)}}$ that capture object attributes, functional roles, and contextual relationships within the surgical environment.

Query Generation with LLM. Given the constructed DT representation $\mathcal{J} = \{\mathcal{J}^{(1)}, \mathcal{J}^{(2)}, \ldots, \mathcal{J}^{(T)}\}$ per video, we then employ an LLM-based agent framework to generate the temporally-constrained RS queries that embed implicit temporal boundaries using $\mathcal{J}$. Specifically, it consists of three sequential stages: (1) identification of the object candidate through ensemble voting, (2) verification of temporal alignment, and (3) generation of queries. The first stage leverages an ensemble voting based on multiple LLMs to identify potential objects and their associated temporal phases from the DT representation for subsequent query generation. Specifically, we deploy LLM instances with different prompts $\{\mathcal{L}_1, \mathcal{L}_2, \ldots, \mathcal{L}_K\}$ to independently evaluate object candidates and their relevance for temporally-constrained video RS task. For each object i identified in the DT representation at time step t, we define a voting function by LLM:

$$v_k^{(i,t)} = \mathcal{L}_k(\mathcal{J}^{(t)}, s_i^{(t)}, \phi_t), \quad (4)$$

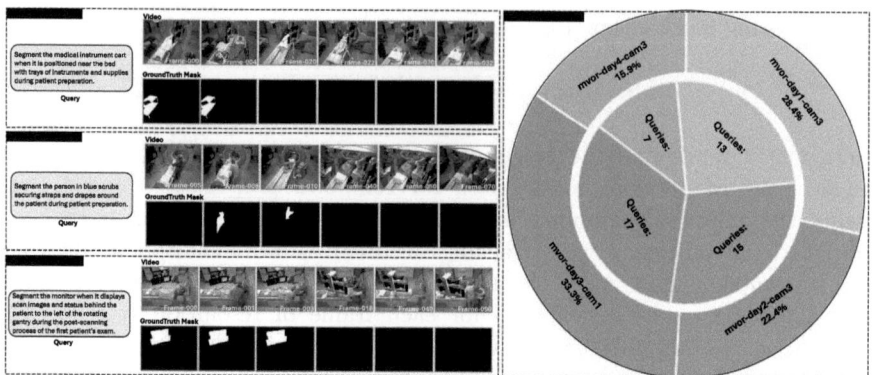

Fig. 3. Overview of *TCVideoRSBenchmark* dataset composition and representative examples of temporally-constrained video RS. (a–c) Three exemplar queries demonstrating temporal constraint reasoning across different procedural phases: (a) segmenting medical instrument cart positioned near the bed during patient preparation (mvor-day1_cam3_04), (b) segmenting personnel in blue scrubs securing patient drapes during patient preparation (mvor-day2_cam3_10), and (c) segmenting monitor displaying scan images during post-scanning process (mvor-day3_cam1_05). Each example includes the temporally-constrained reasoning query, corresponding video frame sequence, and ground truth segmentation masks that are active only during the specified temporal boundaries. (d) Dataset distribution across four MVOR videos, totaling 52 temporally-constrained video RS samples.

where $s_i^{(t)}$ represents the object-level semantic description of object i at time t, and ϕ_t denotes the video-level phase in the corresponding temporal window. Each LLM $\mathcal{L}_k$ assigns a relevance score $v_k^{(i,t)} \in [0,1]$ based on the semantic attributes, spatial configuration, and temporal context of the object within the identified surgical phase. The ensemble voting result for each candidate object-phase pair (i, ϕ) is computed through weighted aggregation $V^{(i,\phi)} = \frac{1}{K} \sum_{k=1}^{K} \sum_{t \in T_\phi} v_k^{(i,t)}$. w_k, where T_ϕ represents the temporal window corresponding to phase ϕ, and w_k denotes the weight assigned to LLM $\mathcal{L}_k$ (where we set $w_k = \frac{1}{K}$ in this work for simplicity). All objects with voting scores exceeding a pre-determined threshold θ_{vote} are selected as candidates for subsequent query generation. Therefore, we can have multiple objects selected from this stage. The second stage performs temporal alignment verification to ensure coherence between selected objects and their associated temporal phases. For each pair of candidates (i^*, ϕ^*) selected from the voting process, we verify temporal consistency through:

$$\tau_{align}(i^*, \phi^*) = \mathbf{1} \left[\exists t \in T_{\phi^*} : \epsilon_i^{(t)} > \theta_{conf} \wedge a_i^{(t)} \neq \emptyset \right], \quad (5)$$

where $\epsilon_i^{(t)}$ represents the confidence score of the object i at time t, $a_i^{(t)}$ denotes the description of the temporal action of the object and θ_{conf} is the confidence threshold. It ensures that selected objects exhibit a meaningful presence and activity during their associated procedural phases. The final stage synthesizes temporally-constrained video RS queries by embedding implicit temporal boundaries within natural language formulations. Given a validated object-phase pair (i^*, ϕ^*), we construct the reasoning query through template-based generation:

$$Q_{temp} = \mathcal{G}(s_{i^*}, \phi^*, T_{\phi^*}, \mathcal{R}_{spatial}, \mathcal{R}_{semantic}), \quad (6)$$

where $\mathcal{G}$ represents the LLM for query generation, $\mathcal{R}_{spatial}$ and $\mathcal{R}_{semantic}$ denote the spatial and semantic relationship descriptors extracted from the DT representation. The corresponding ground truth RS masks are extracted from the DT representation by applying temporal constraints to the pre-existing instance masks. Specifically, for the selected object i^* and phase ϕ^*, the temporally-constrained ground truth is constructed as $\mathcal{M}_{gt} = \{M_{i^*}^{(t)} \cdot \tau_{\phi^*}(t) | t = 1, 2, \ldots, T\}$, where $M_{i^*}^{(t)}$ represents the instance mask for object i^* at time t stored in the DT representation, and $\tau_{\phi^*}(t)$ is the temporal constraint function that equals 1 when $t \in T_{\phi^*}$ and 0 otherwise. Finally, all the generated samples (the triplet query, video, and masks) go through manual verification, and we filter the incorrect or improper ones.

Dataset Statistics. The video sequences in TCVideoRSBenchmark are from the MVOR dataset [12], which is an authentic OR dataset, consisting of 732 synchronized frames at 640 × 480 resolution. TCVideoRSBenchmark comprises 52 temporally-constrained video RS samples from 4 representative videos, with the following distribution mvor-day1_cam3 contributes 15 queries (28.4%), mvor-day2_cam3 provides 13 queries (22.4%), mvor-day3_cam1 contains 17 queries (33.3%), and mvor-day4_cam3 includes 7 queries (15.9%), as illustrated in Fig. 3. Each sample in *TCVideoRSBenchmark* consists of a temporally-constrained reasoning query paired with video and corresponding ground truth segmentation masks that are temporally bounded to specific procedural phases. The queries encompass diverse surgical workflow scenarios, where representative examples in Fig. 3 include segmenting medical instrument carts during patient preparation phases. The temporal constraints embedded within the queries span multiple procedural phases commonly observed in operating room workflows, including patient preparation, equipment setup, active intervention periods, and post-procedural activities.

4 Conclusion

The innovative formulation of temporally-constrained video RS aims to address the conventional RS's limitation of assuming continuous object tracking. As a new task, we propose an automated benchmark construction pipeline that uses DT representations, demonstrating the potential to create scalable benchmark datasets for temporally-constrained video RS without manual annotation efforts. Based on this method, the *TCVideoRSBenchmark*, including 52 samples derived from the MVOR dataset, is constructed to allow evaluation of the ability to understand when segmentation objectives become contextually applicable within surgical procedures. Overall, the demonstration of temporally-constrained video RS opens new possibilities for surgical workflow monitoring methods that provide contextually relevant insights aligned with the natural temporal structure of medical procedures.

Acknowledgments. This work was supported in part by the JHU Amazon Initiative for Artificial Intelligence (AI2AI) fellowship program and NSF CAREER award (NSF Award No. 2239077).

References

1. Bai, S., et al.: Qwen2. 5-VL technical report. arXiv preprint arXiv:2502.13923 (2025)
2. Demir, K.C., Rodriguez, B.O., Weise, T., Maier, A., Yang, S.H.: Towards intelligent speech assistants in operating rooms: a multimodal model for surgical workflow analysis. arXiv preprint arXiv:2406.14576 (2024)
3. Grammatikopoulou, M., et al.: A spatio-temporal network for video semantic segmentation in surgical videos. Int. J. Comput. Assist. Radiol. Surg. **19**(2), 375–382 (2024)
4. Jin, Y., Long, Y., Chen, C., Zhao, Z., Dou, Q., Heng, P.-A.: Temporal memory relation network for workflow recognition from surgical video. IEEE Trans. Med. Imaging **40**(7), 1911–1923 (2021)
5. Lai, X., Tian, Z., Chen, Y., et al.: LISA: reasoning segmentation via large language model. In: Proceedings of the IEEE/CVF Conference on Computer Vision and Pattern Recognition, pp. 9579–9589 (2024)
6. Liu, H., Li, C., Li, Y., Lee, Y.J.: Improved baselines with visual instruction tuning. In: Proceedings of the IEEE/CVF Conference on Computer Vision and Pattern Recognition, pp. 26296–26306 (2024)
7. Ravi, N., Gabeur, V., Hu, Y.-T., et al.: SAM 2: segment anything in images and videos. arXiv preprint arXiv:2408.00714 (2024)
8. Shen, Y., Ding, H., Seenivasan, L., Shu, T., Unberath, M.: Position: foundation models need digital twin representations. arXiv preprint arXiv:2505.03798 (2025)
9. Shen, Y., Li, C., Fan, C., Unberath, M.: RVTBench: a benchmark for visual reasoning tasks. arXiv preprint arXiv:2505.11838 (2025)
10. Shen, Y., Li, C., Liu, B., Li, C.-Y., Porras, T., Unberath, M.: Operating room workflow analysis via reasoning segmentation over digital twins. arXiv preprint arXiv:2503.21054 (2025)
11. Shen, Y., et al.: Reasoning segmentation for images and videos: a survey. arXiv preprint arXiv:2505.18816 (2025)
12. Srivastav, V., Issenhuth, T., Kadkhodamohammadi, A., de Mathelin, M., Gangi, A., Padoy, N.: MVOR: a multi-view RGB-D operating room dataset for 2D and 3D human pose estimation. arXiv preprint arXiv:1808.08180 (2018)
13. Xu, J., Sirajudeen, N., Boal, M., Francis, N., Stoyanov, D., Mazomenos, E.B.: SedMamba: enhancing selective state space modelling with bottleneck mechanism and fine-to-coarse temporal fusion for efficient error detection in robot-assisted surgery. IEEE Robot. Autom. Lett. (2024)
14. Yang, L., Kang, B., Huang, Z., et al.: Depth anything v2. arXiv preprint arXiv:2406.09414 (2024)

Cross-Modal Knowledge Distillation for Chest Radiographic Diagnosis via Embedding Expansion, Reconstruction, and Classification

DongAo Ma[1], Jiaxuan Pang[1], Ziyu Zhou[2], Michael B. Gotway[3], and Jianming Liang[1]($\boxtimes$)

[1] Arizona State University, Tempe, AZ 85281, USA
{dongaoma,jpang12,jianming.liang}@asu.edu
[2] Shanghai Jiao Tong University, Minhang, Shanghai 200240, China
zhouziyu@sjtu.edu.cn
[3] Mayo Clinic, Scottsdale, AZ 85259, USA
Gotway.Michael@mayo.edu

Abstract. Deep learning models now can achieve expert-level performance for chest radiographic (CXR) diagnosis, but they require pretraining with large-scale image datasets and categorical labels. Such labels are, however, frequently incomplete and commonly fail to capture the full spectrum of clinical knowledge embedded in radiology reports issued by expert radiologists. Large language models (LLMs) can encode text into machine-understandable semantic representations, providing an opportunity for cross-modal knowledge distillation, thereby enhancing vision-based CXR diagnosis models. We propose **CREED**—**C**lassifying and **R**econstructing **E**xpanded **E**mbeddings for cross-modal knowledge **D**istillation to address this shortcoming. CREED has three learning objectives: (1) *embedding reconstruction* to preserve fine-grained language information, (2) *diagnostic classification* to bridge the modality gap, and (3) *KL-divergence minimization* to enforce alignment between vision and language embeddings. Extensive experiments across diverse diagnostic tasks and on report generation demonstrate that CREED consistently outperforms baseline and state-of-the-art (SOTA) fully-/self-supervised models. The results highlight the effectiveness of cross-modal knowledge distillation from LLMs for enhancing CXR diagnosis by infusing clinically rich semantics into vision models, ultimately improving diagnostic accuracy, interpretability, and generalization across diverse medical imaging tasks. The codes and pretrained models are available at GitHub.com/JLiangLab/CREED.

Keywords: Cross-Modal Knowledge Distillation · Chest Radiography

1 Introduction

Automated chest radiographic (CXR) analysis is rapidly becoming a crucial tool in medical imaging, leveraging deep learning models to assist radiologists in detecting and classifying thoracic diseases [1,10,14]. While these models have demonstrated expert-level performance, they predominantly rely on categorical labels, which are often incomplete and fail to capture the full spectrum of clinical knowledge embedded in reports issued by expert radiologists [5,15,17–19,22] (see Fig. 1). As a result, these models may lack the nuanced understanding required for appropriate medical decision-making [29]. The advancement of LLMs presents a promising opportunity to bridge this gap, because LLMs can effectively process and summarize expert knowledge within radiology reports, encoding them into machine-understandable semantic representations (*i.e.* language embeddings, which are information-rich numerical vectors capturing clinical semantics). This capability enables the potential for cross-modal knowledge distillation, transferring detailed clinical information within reports via LLMs to vision-based models. However, a significant challenge arises due to the substantial domain gap between textual language embeddings contained within reports and visual feature embeddings from images, making direct alignment difficult. To address this critical limitation, this work explores a critical question: *How can we effectively distill detailed clinical semantic knowledge embedded in radiology reports via LLMs into vision-based CXR diagnosis models?*

We propose **CREED**, a framework that **C**lassifies and **R**econstructs **E**xpanded **E**mbeddings for cross-modal knowledge **D**istillation, as illustrated in Fig. 2, enabling the transfer of clinically relevant knowledge with detailed semantics from LLMs into vision-based CXR diagnosis models. CREED distills rich clinical semantics from LLM-generated embeddings into a vision model through three learning objectives: (1) *embedding reconstruction* to preserve fine-grained language information, (2) *diagnostic classification* to bridge the modality gap, and (3) *KL-divergence minimization* to enforce alignment between vision and language embeddings. We evaluated CREED across a variety of diagnostic tasks, including long-tailed, multi-class, multi-label, and binary classification (Table 1), and demonstrated that CREED consistently outperforms or matches state-of-the-art (SOTA) supervised and self-supervised models. Our results highlight CREED's effectiveness for enhancing CXR diagnosis through cross-modal knowledge transfer. Additionally, CREED improves performance in radiology report generation (Table 2), showcasing its ability to learn detailed clinical semantic knowledge from LLMs.

Unlike exisiting visual-language pretraining models that align image and text representations using paired medical data (e.g., CXRs and radiology reports) through contrastive learning objectives [3,23,28], or the approach that incorporate medical knowledge graphs to encode structured pathology relationships [30], CREED distills knowledge from embeddings generated by frozen LLMs based on radiology reports, utilizing embedding reconstruction alongside a classification task with diagnostic labels to bridge the domain gap effectively.

Through this work, we have made the following contributions: (1) A model that distills clinical knowledge generated by LLMs based on radiology reports; (2) A framework, comprising an encoder, decoder, and classifier, shared between the LLM and the vision model, which expands, reconstructs, and classifies the embeddings to align the language and vision domains; (3) Comprehensive experiments that evaluate CREED on various diagnostic tasks and report generation, demonstrating its enhanced capability for CXR diagnosis.

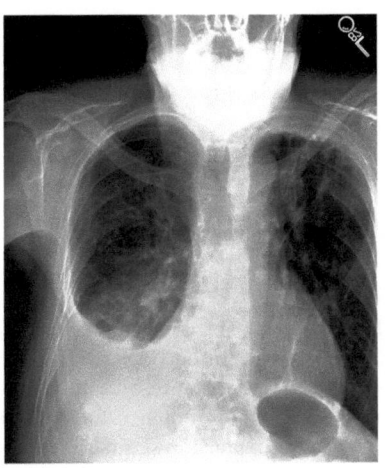

Fig. 1. Conventional training for chest radiographic diagnosis models primarily relies on structured labels (right bottom table), which are often incomplete, overlooking the rich clinical information embedded in reports created by expert radiologists. Radiology reports provide detailed descriptions of abnormalities, their severity, and contextual findings (highlighted with colors) that are not explicitly captured by categorical labels. Our approach distills essential knowledge contained within radiology reports into the vision model, enhancing its ability to comprehend nuanced diagnostic patterns beyond predefined labels, ultimately improving diagnostic performance.

2 Method

Architectural Components. The CREED framework is designed to distill knowledge from an LLM into a vision model by leveraging CXR images and their corresponding reports. At its core, CREED introduces a shared framework that operates between the LLM and the vision model, consisting of three key components: encoder, decoder, and classifier. The encoder expands input embeddings into a higher-dimensional latent space, in contrast to conventional autoencoders that compresses inputs. This expansion aligns embeddings from vision and language domains into a shared latent space, effectively integrating textual and visual semantics to enable cross-modal knowledge distillation. The intuition behind this design is to construct representations that fuse vision–language mutual information within a richer, higher-dimensional space. The decoder reconstructs the input language embeddings from the expanded latent

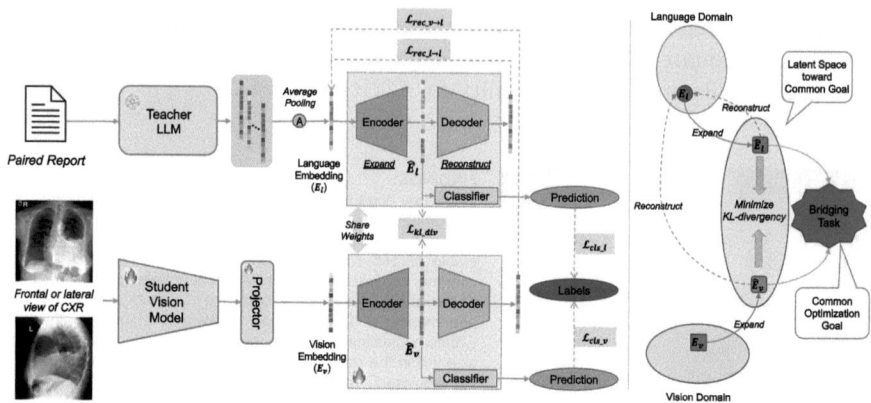

Fig. 2. CREED framework distills knowledge embedded in radiology reports via a *frozen* teacher LLM into a student vision model. Designed as a framework *shared* between the LLM and the vision model, CREED consists of an encoder, a decoder, and a linear classifier. Unlike conventional autoencoders that compress input into a lower-dimensional latent space, the CREED encoder expands input embeddings into a higher dimension. When processing vision embeddings, this expansion should integrate knowledge from the language domain, enhancing representation learning. To achieve this, we introduce three learning objectives that bridge vision and language embeddings: a reconstruction task, a classification task, and KL-divergence minimization. The reconstruction task employs a decoder to recover the original language embedding from the expanded language or vision embeddings ($\mathcal{L}_{rec\_l \to l} + \mathcal{L}_{rec\_v \to l}$). By requiring the expanded vision embedding to reconstruct the language embedding, CREED is forced to distill essential knowledge from the LLM. To facilitate cross-modal distillation between heterogeneous models and data, we introduce a bridging classification task as a common optimization goal across both modalities. By using diagnostic labels ($\mathcal{L}_{cls\_l} + \mathcal{L}_{cls\_v}$), this task aligns input embeddings from different modalities into a shared latent space. Additionally, we further enforce similarity between the latent spaces by minimizing the KL-divergence between the two expanded embeddings ($\mathcal{L}_{kl-div}$). The conceptual framework of CREED is illustrated on the right side.

space with either language or vision embeddings, ensuring knowledge transfer from the LLM to the vision model. The classifier maps vision and language embeddings into a shared diagnostic label space, unifying heterogeneous modalities through a common optimization objective.

Learning Objectives. To optimize cross-modal distillation, we introduce three learning objectives that bridge vision and language embeddings:

Embedding Reconstruction Task. The decoder reconstructs language embeddings from both modalities, facilitating knowledge distillation from the LLM to the vision model:

$$\mathcal{L}_{rec} = \mathcal{L}_{rec\_l \to l}(E_l, Decoder(\widehat{E}_l)) + \mathcal{L}_{rec\_v \to l}(E_l, Decoder(\widehat{E}_v)) \quad (1)$$

where $\mathcal{L}_{rec\_l \to l}$ and $\mathcal{L}_{rec\_v \to l}$ correspond to language-to-language and vision-to-language embedding reconstruction losses, respectively. Here, E_l represents the original language embedding from the LLM, while $\widehat{E}_l$ and $\widehat{E}_v$ denote the expanded language and vision embeddings, respectively.

Bridging Classification Task. A shared classifier aligns diagnostic predictions across modalities using labels provided with the dataset, facilitating the vision and language embeddings to be mapped into a shared latent space:

$$\mathcal{L}_{cls} = \mathcal{L}_{cls\_l}(y, Classifier(\widehat{E}_l)) + \mathcal{L}_{cls\_v}(y, Classifier(\widehat{E}_v)) \tag{2}$$

KL-Divergence Minimization. To reduce distributional discrepancies between expanded vision and language embeddings, we applied:

$$\mathcal{L}_{kl-div} = D_{KL}(\widehat{E}_v \parallel \widehat{E}_l) \tag{3}$$

to ensure statistical similarity between modalities, complementing the embedding reconstruction task.

The overall training objective integrates reconstruction, classification, and distribution alignment losses into a unified optimization framework:

$$\mathcal{L} = \mathcal{L}_{rec} + \mathcal{L}_{cls} + \mathcal{L}_{kl-div} \tag{4}$$

By expanding embeddings rather than compressing them, CREED retains critical clinical semantics contained within radiology reports. The synergy of reconstruction, classification, and distribution alignment enables robust cross-modal knowledge transfer, empowering the vision model to outperform purely visual approaches that only learn from labels.

Implementation Details. The frozen Llama-3.1-8B [4] model serves as the teacher LLM, initialized with pretrained weights from the Meta Llama repository. For the student model, we adopted a Swin-Base Transformer (Swin-B) [11] pretrained on ImageNet-22k, using an input resolution of 224 × 224. We replaced the final classification layer with a projector to align the output dimension with Llama's hidden representation while retaining its hierarchical feature extraction capabilities. The encoder and decoder are implemented using multilayer perceptrons (MLPs), whereas the classifier is a single linear layer. We employed mean squared error (MSE) loss to enforce embedding reconstruction and binary cross-entropy (BCE) loss for multi-label classification for training objectives.

We utilized the official training split of the MIMIC-CXR dataset [8], consisting of 354,538 CXR images paired with their corresponding radiology reports, for pretraining. We extracted and processed the "Findings" and "Impression" sections of the reports, as they encapsulate radiologists' diagnostic conclusions, providing rich textual inputs for the LLM. Additionally, we adopted the 14 structured labels officially released with the dataset (derived from the radiology reports) as ground-truth targets for cross-modal alignment.

3 Experiments and Results

3.1 CREED Achieves SOTA Performance in Various Diagnostic Tasks

Experimental Setup. We conducted extensive experiments across diverse diagnostic tasks, including long-tailed, multi-class, multi-label, and binary classification to evaluate CREED. We benchmarked CREED against four comparative models: (1) a supervised baseline using an ImageNet-pretrained model [11], fine-tuned on the benchmark datasets; (2) the SOTA fully-supervised Ark-6 [13] model trained on 704K CXRs; (3) SimMIM-CXR [12]; and (4) RAD-DINO [21], both self-supervised models pretrained on 926K and 882K CXRs, respectively. RAD-DINO adopts a ViT-B backbone, whereas others use Swin-B. Downstream models initialized their encoders with pretrained weights and reinitialized task-specific classification heads according to target class number to ensure fair comparison. All layers were fine-tuned under identical experimental protocols following [12]. The experimental details for fine-tuning are available at GitHub.com/JLiangLab/BenchmarkTransformers.

We employ four datasets to comprehensively evaluate CREED's performance: ChestDR [25], which exhibits a long-tailed class distribution comprising 10 head classes and 9 tail classes; COVIDx [26], designed to distinguish among patients without pneumonia, with non-COVID-19 pneumonia, and with COVID-19 pneumonia cases; VinDrCXR [16], featuring six image-level global labels for multi-label classification; and ShenzhenCXR [6], focusing on screening tuberculosis (TB) from normal CXRs. Metrics used to assess model performance included AUC, Accuracy (ACC), F1 Score, Average Precision (AP), and Matthews Correlation Coefficient (MCC). Results are reported as means ± standard deviations over five independent runs on hold-out test sets. Statistical significance is assessed via independent two-sample t-tests ($p < 0.05$).

Results and Analysis. Table 1 presents the performance comparison between CREED and existing SOTA models across various thoracic disease diagnostic tasks. CREED achieves a mean AUC of 82.70% when fine-tuned at 224×224 resolution on ChestDR, significantly surpassing the ImageNet-pretrained baseline and the self-supervised SimMIM-CXR model, while performing comparably to the SOTA supervised Ark-6 and self-supervised RAD-DINO models. Notably, when fine-tuned at a higher resolution of 448×448, CREED attains the highest AUC (83.38%), MCC (31.36%), AP (38.57%), and F1 score (34.14%), outperforming both Ark-6 and RAD-DINO. These results highlight CREED's strong capacity for enhancing representation learning in long-tailed classification tasks. Additionally, CREED achieves the highest accuracy (97.75%) and MCC (96.89%) on COVIDx. CREED also demonstrates statistically comparable performance to Ark-6 on VinDrCXR and ShenzhenCXR, even when the latter had the advantage of incorporating training images and labels from these datasets during pretraining. These results collectively underscore CREED's effectiveness for vision-based chest radiographic diagnosis across diverse classification tasks,

Table 1. CREED achieves SOTA performance across various thoracic disease diagnostic tasks, outperforming or matching existing SOTA fully- and self-supervised models, even those pretrained on larger multi-source datasets. This underscores the CREED's effectiveness for achieving superior diagnostic accuracy, demonstrating the impact of cross-modal distillation from LLMs and radiology reports. Performance values are formatted as follows: bold for the best result, underlined for the second-best, and green shading to indicate methods with no statistically significant difference ($p \geq 0.05$) from the best result.

Method	Input resolution	ChestDR AUC (%)	MCC (%)	AP (%)	F1 (%)
Baseline		77.39±0.48	25.19±0.52	28.41±0.62	29.34±0.39
SimMIM-CXR	224×224	78.03±0.95	25.25±0.92	28.06±1.01	28.99±0.58
Ark-6		82.58±0.12	31.01±0.21	37.03±0.36	33.82±0.26
CREED		82.70±0.10	30.53±0.58	37.05±0.85	33.32±0.70
RAD-DINO	518×518	82.73±0.57	31.10±1.16	37.24±1.90	33.89±1.11
CREED*	448×448	**83.38±0.15**	**31.36±0.28**	**38.57±0.80**	**34.14±0.38**

Method	Input resolution	COVIDx Accuracy (%)	MCC (%)	VinDrCXR AUC (%)	ShenzhenCXR AUC (%)
Baseline		96.10±0.83	92.28±0.33	90.35±0.31	93.35±0.77
SimMIM-CXR	224×224	96.17±0.56	93.88±2.02	91.71±1.04	95.76±1.79
Ark-6†		97.25±0.40	95.34±0.57	**95.07±0.16**	**98.98±0.15**
CREED		97.40±0.34	95.30±0.80	94.06±0.18	98.28±0.36
RAD-DINO‡	518×518	97.05±0.27	95.58±0.81	94.74±0.35	98.66±0.08
CREED*	448×448	**97.75±0.64**	**96.89±0.54**	94.64±0.55	98.92±0.35

* Our experiments use the same pretrained model (initialized at 224×224 resolution), with input resolution increased to 448×448 during fine-tuning.
† Ark-6 incorporated training images and labels from the VinDrCXR and ShenzhenCXR datasets during its pretraining phase, which may confer a comparative advantage in downstream tasks.
‡ RAD-DINO, continuously trained from DINOv2-Base pretrained on LVD-142M (a dataset vastly larger than ImageNet), leverages more CXR images and higher-resolution inputs for self-supervised pretraining, positioning it as a strong foundation model for comparison.

further highlighting its capability to transfer knowledge from LLMs for robust cross-modal representation learning.

3.2 CREED Boosts Performance for Radiology Report Generation

Experimental Setup. Automated radiology report generation has become a sophisticated yet highly valuable AI-driven task due to the advancement of LLMs. By generating coherent paragraphs that accurately capture the observations and findings from radiology images, this technology can significantly reduce radiologists' workload and diagnostic errors, accelerating clinical workflows [2,27]. To demonstrate that CREED effectively distills detailed clinical knowledge from LLM embeddings into its vision model, we evaluate the model on radiology report generation using the MIMIC-CXR [8] dataset, following the framework and configuration of R2GenGPT [27]. R2GenGPT employs a visual alignment module, implemented as a linear layer, to map visual features from a visual model into the word embedding space of a frozen LLM, enabling the generation of coherent reports from radiographic images. We adopted the Shallow Alignment configuration to isolate the impact of knowledge distilled from the LLM during CREED's pretraining, training only the linear alignment layer while keeping the vision model fixed. We also evaluated CREED under the Deep Alignment setup, where the entire model was fine-tuned. In both settings, the vision model was initialized with pretrained weights from CREED, ensuring a direct evaluation of its effectiveness in report generation. We employed several metrics to assess the quality of generated reports against radiologists' written reports, including BLEU scores [20], ROUGE-L [9], and CIDEr [24], according to the established evaluation protocol [27].

Table 2. CREED improves report generation performance under both R2GenGPT [27] configurations. Notably, in the Shallow Alignment setting, with only 4.2M trainable parameters, CREED achieves performance comparable to the Deep Alignment baseline, which uses 90.9M trainable parameters.

Configuration	Param.	Model	Bleu-1	Bleu-2	Bleu-3	Bleu-4	ROUGE-L	CIDEr
Shallow Alignment	4.2M	R2GenGPT	0.364	0.235	0.161	0.116	0.276	0.135
		CREED	0.401	0.261	0.182	0.130	0.296	0.195
Deep Alignment	90.9M	R2GenGPT	0.406	0.268	0.186	0.135	0.298	0.262
		CREED	0.413	0.269	0.189	0.139	0.304	0.268

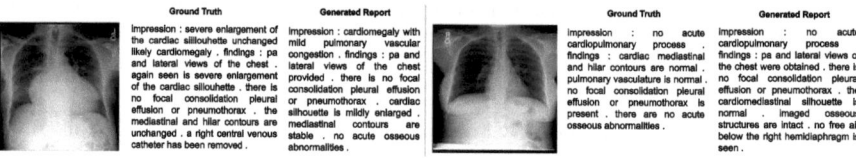

Fig. 3. Examples of radiology reports generated by R2GenGPT using CREED. Key medical findings are highlighted with colors for better illustration.

Results and Analysis. Table 2 presents the comparison between the baseline R2GenGPT model and CREED across multiple text generation metrics, including BLEU-n, ROUGE-L, and CIDEr. CREED improves radiology report generation performance across all evaluated metrics under both configurations. Under the Shallow Alignment configuration, the BLEU-1 to BLEU-4 scores increase by 0.037, 0.026, 0.020, and 0.014, respectively, indicating enhanced unigram (word-level) precision and phrase-level coherence in the generated reports. ROUGE-L rises from 0.276 to 0.296, reflecting improved semantic continuity and structural alignment with human-written reports. Notably, CIDEr demonstrates a substantial improvement, increasing from 0.135 to 0.195, suggesting that CREED generates reports that are more informative, contextually relevant, and diverse. Deep Alignment can further improve the performance, especially for the CIDEr metric. Additionally, Fig. 3 illustrates two example reports generated using CREED, with key medical findings highlighted to demonstrate their alignment with expert-written reports. The results underscore the effectiveness of CREED's cross-modal knowledge distillation, demonstrating its ability to bridge the gap between visual features and language embeddings, leading to improved performance in automated radiology report generation, enabling more accurate and coherent medical descriptions.

Table 3. Performance degradation caused by ablating key components.

Configuration	ChestDR (AUC%)	ShenzhenCXR (AUC%)
Full model (CREED)	82.70 ± 0.10	98.28 ± 0.36
No reconstruction loss	81.32 ± 0.51 (↓1.38)	97.43 ± 0.33 (↓0.85)
No classification loss	80.04 ± 0.75 (↓2.66)	96.31 ± 0.45 (↓1.97)
No KL-divergence loss	81.42 ± 0.43 (↓1.28)	97.09 ± 0.03 (↓1.19)

3.3 Ablation Study

We systematically ablate the reconstruction loss ($\mathcal{L}_{rec}$), classification loss ($\mathcal{L}_{cls}$), and KL-divergence loss ($\mathcal{L}_{kl-div}$) to quantify the contribution of each loss component in CREED. The results on ChestDR and ShenzhenCXR are summarized in Table 3. Removing any of these losses leads to a significant performance decrement, underscoring their complementary roles in effective cross-modal distillation. Notably, the reconstruction and classification losses work synergistically: reconstruction preserves fine-grained semantics, whereas classification ensures task-relevant alignment, reinforcing CREED's robustness.

4 Conclusion and Future Work

We have introduced CREED, a novel cross-modal knowledge distillation framework that integrates radiology report semantics into vision-based CXR diagnosis models using embeddings from LLMs. CREED bridges the modality gap between vision and language by employing embedding reconstruction, diagnostic classification, and KL-divergence minimization. Our experiments show that CREED achieves SOTA performance across various diagnostic tasks and enhances report generation.

Future work will focus on incorporating additional datasets for large-scale pretraining, leveraging advanced medical LLMs as teacher models, and introducing explicit multimodal interaction mechanisms (*e.g.* attention modules, fusion layers, and token-level correspondence) to further enhance model performance. Additionally, a benchmark needs to be established for evaluating the model's ability to assess diagnostic uncertainty and disease severity, and to validate its effectiveness on real-world clinical cases.

Acknowledgements. This research has been supported in part by ASU and Mayo Clinic through a Seed Grant and an Innovation Grant, and in part by the NIH under Award Number R01HL128785. The content is solely the responsibility of the authors and does not necessarily represent the official views of the NIH. This work has utilized the GPUs provided in part by the ASU Research Computing [7] and in part by the Bridges-2 at Pittsburgh Supercomputing Center and the Anvil at Purdue University through allocation MED220025 from the Advanced Cyberinfrastructure Coordination Ecosystem: Services & Support (ACCESS) program, which is supported by National Science Foundation grants #2138259, #2138286, #2138307, #2137603, and #2138296.

References

1. Çallı, E., Sogancioglu, E., van Ginneken, B., van Leeuwen, K.G., Murphy, K.: Deep learning for chest X-ray analysis: a survey. Med. Image Anal. **72**, 102125 (2021)
2. Chen, Z., Song, Y., Chang, T.H., Wan, X.: Generating radiology reports via memory-driven transformer. arXiv preprint arXiv:2010.16056 (2020)
3. Dai, T., Zhang, R., Hong, F., Yao, J., Zhang, Y., Wang, Y.: UniChest: conquer-and-divide pre-training for multi-source chest X-ray classification. IEEE Trans. Med. Imaging (2024)
4. Dubey, A., et al.: The LLaMA 3 herd of models. arXiv preprint arXiv:2407.21783 (2024)
5. Islam, N.U., Ma, D., Pang, J., Velan, S.S., Gotway, M.B., Liang, J.: Foundation X: integrating classification, localization, and segmentation through lock-release pretraining strategy for chest X-ray analysis. In: Proceedings of the IEEE/CVF Winter Conference on Applications of Computer Vision (2025)
6. Jaeger, S., Candemir, S., Antani, S., Wáng, Y.X.J., Lu, P.X., Thoma, G.: Two public chest X-ray datasets for computer-aided screening of pulmonary diseases. Quant. Imaging Med. Surg. **4**(6), 475 (2014)
7. Jennewein, D.M., et al.: The sol supercomputer at Arizona state university. In: Practice and Experience in Advanced Research Computing, pp. 296–301 (2023)
8. Johnson, A.E., et al.: MIMIC-CXR, a de-identified publicly available database of chest radiographs with free-text reports. Sci. data **6**(1), 1–8 (2019)
9. Lin, C.Y.: ROUGE: a package for automatic evaluation of summaries. In: Text Summarization Branches Out, pp. 74–81 (2004)
10. Litjens, G., et al.: A survey on deep learning in medical image analysis. Med. Image Anal. **42**, 60–88 (2017)

11. Liu, Z., et al.: Swin transformer: hierarchical vision transformer using shifted windows. In: Proceedings of the IEEE/CVF International Conference on Computer Vision, pp. 10012–10022 (2021)
12. Ma, D., et al.: Benchmarking and boosting transformers for medical image classification. In: Kamnitsas, K., et al. (eds.) DART 2022. LNCS, vol. 13542, pp. 12–22. Springer, Cham (2022). https://doi.org/10.1007/978-3-031-16852-9_2
13. Ma, D., Pang, J., Gotway, M.B., Liang, J.: Foundation ark: accruing and reusing knowledge for superior and robust performance. In: Greenspan, H., et al. (eds.) MICCAI 2023. LNCS, vol. 14220, pp. 651–662. Springer, Cham (2023). https://doi.org/10.1007/978-3-031-43907-0_62
14. Ma, D., Pang, J., Gotway, M.B., Liang, J.: A fully open AI foundation model applied to chest radiography. Nature 1–11 (2025). https://doi.org/10.1038/s41586-025-09079-8
15. Ma, D., Pang, J., Velan, S.S., Gotway, M.B., Liang, J.: ARK+: accruing and reusing knowledge from heterogeneous labels across numerous datasets for training superior and robust AI foundation models. Med. Image Anal. (2025). (to appear)
16. Nguyen, H.Q., et al.: VinDr-CXR: an open dataset of chest X-rays with radiologist's annotations. Sci. Data **9**(1), 429 (2022)
17. Pang, J., et al.: POPAR: patch order prediction and appearance recovery for self-supervised medical image analysis. In: Kamnitsas, K., et al. (eds.) DART 2022. LNCS, vol. 13542, pp. 77–87. Springer, Cham (2022). https://doi.org/10.1007/978-3-031-16852-9_8
18. Pang, J., Ma, D., Zhou, Z., Gotway, M.B., Liang, J.: ASA: learning anatomical consistency, sub-volume spatial relationships and fine-grained appearance for CT images. In: Linguraru, M.G., et al. (eds.) MICCAI 2024. LNCS, vol. 15011, pp. 91–101. Springer, Cham (2024). https://doi.org/10.1007/978-3-031-72120-5_9
19. Pang, J., Ma, D., Zhou, Z., Gotway, M.B., Liang, J.: POPAR: patch order prediction and appearance recovery for self-supervised learning in chest radiography. Med. Image Anal. 103720 (2025)
20. Papineni, K., Roukos, S., Ward, T., Zhu, W.J.: BLEU: a method for automatic evaluation of machine translation. In: Proceedings of the 40th Annual Meeting of the Association for Computational Linguistics, pp. 311–318 (2002)
21. Pérez-García, F., et al.: RAD-DINO: exploring scalable medical image encoders beyond text supervision (2024)
22. Sellergren, A.B., et al.: Simplified transfer learning for chest radiography models using less data. Radiology **305**(2), 454–465 (2022)
23. Tiu, E., Talius, E., Patel, P., Langlotz, C.P., Ng, A.Y., Rajpurkar, P.: Expert-level detection of pathologies from unannotated chest X-ray images via self-supervised learning. Nat. Biomed. Eng. **6**(12), 1399–1406 (2022)
24. Vedantam, R., Lawrence Zitnick, C., Parikh, D.: CIDEr: consensus-based image description evaluation. In: Proceedings of the IEEE Conference on Computer Vision and Pattern Recognition, pp. 4566–4575 (2015)
25. Wang, D., et al.: A real-world dataset and benchmark for foundation model adaptation in medical image classification. Sci. Data **10**(1), 574 (2023)
26. Wang, L., Lin, Z.Q., Wong, A.: COVID-net: a tailored deep convolutional neural network design for detection of COVID-19 cases from chest x-ray images. Sci. Rep. **10**(1), 19549 (2020)
27. Wang, Z., Liu, L., Wang, L., Zhou, L.: R2GenGPT: radiology report generation with frozen LLMs. Meta-Radiol. **1**(3), 100033 (2023)

28. Wang, Z., Wu, Z., Agarwal, D., Sun, J.: MedCLIP: contrastive learning from unpaired medical images and text. In: Proceedings of the Conference on Empirical Methods in Natural Language Processing. Conference on Empirical Methods in Natural Language Processing, vol. 2022, p. 3876 (2022)
29. Zhang, M., et al.: A new benchmark: clinical uncertainty and severity aware labeled chest X-ray images with multi-relationship graph learning. IEEE Trans. Med. Imaging (2024)
30. Zhang, X., Wu, C., Zhang, Y., Xie, W., Wang, Y.: Knowledge-enhanced visual-language pre-training on chest radiology images. Nat. Commun. **14**(1), 4542 (2023)

Random Direct Preference Optimization for Radiography Report Generation

Valentin Samokhin[1,3], Boris Shirokikh[1,2(✉)], Mikhail Goncharov[1,2], Dmitriy Umerenkov[1,2], Maksim Bobrin[1,2], Ivan Oseledets[1,2], Dmitry Dylov[1,2], and Mikhail Belyaev[4]

[1] Artificial Intelligence Research Institute (AIRI), Moscow, Russia
[2] Skolkovo Institute of Science and Technology, Moscow, Russia
boris.shirokikh@skoltech.ru
[3] Institute for Information Transmission Problems, Moscow, Russia
[4] Moscow, Russia

Abstract. Radiography Report Generation (RRG) has gained significant attention in medical image analysis as a promising tool for alleviating the growing workload of radiologists. However, despite numerous advancements, existing methods have yet to achieve the quality required for deployment in real-world clinical settings. Meanwhile, large Visual Language Models (VLMs) have demonstrated remarkable progress in the general domain by adopting training strategies originally designed for Large Language Models (LLMs), such as alignment techniques. In this paper, we introduce a model-agnostic framework to enhance RRG accuracy using Direct Preference Optimization (DPO). Our approach leverages random contrastive sampling to construct training pairs, eliminating the need for reward models or human preference annotations. Experiments on supplementing three state-of-the-art models with our Random DPO show that our method improves clinical performance metrics by up to 5%, without requiring any additional training data.

Keywords: Radiography · Report Generation · Visual Language Models · Alignment · Direct Preference Optimization

1 Introduction

Chest radiography (CXR) has become the global standard for the non-invasive assessment of major thoracic organs. It is a fast and a widely used screening tool that provides valuable diagnostic insights while minimizing radiation exposure. CXR plays a crucial role in detecting both acute and chronic conditions affecting the lungs, great vessels, mediastinum, and osseous structures [11].

The growing global demand for CXR examinations has placed an increasing burden on radiologists responsible for interpreting these studies. And the number

M. Belyaev—Independent Researcher.

of radiographic studies is rising faster than the number of radiologists available to review them [11]. This increasing workload heightens the risk of diagnostic errors due to fatigue and time constraints. In the context of thoracic pathologies, one of the leading causes of mortality, such errors can have severe consequences.

Automated methods for radiology report generation (RRG), particularly those based on large Visual Language Models (VLMs) [1,3,16], offer a promising solution to alleviating the increasing burden on radiologists. These systems can serve as a second opinion, assisting in the detection of major pathologies and reducing the likelihood of missed findings. However, despite recent advancements, even state-of-the-art solutions fall short of the quality required for deployment in real-world clinical settings [26,30].

Recent studies [1,4,17] have shown that incorporating auxiliary tasks, such as grounding or visual question answering, into the training pipeline enhances model performance on RRG benchmarks. This suggests that the conventional next-token prediction objective, commonly used for radiograph captioning, is insufficient for learning clinically accurate reports. One of the key challenges here is the inherent variability in clinical reports – they are written by different physicians, each with their own preferred style and terminology.

The latter problem can be approached using the general domain methods, where VLMs and Large Language Models (LLMs) ensure their outputs align with highly-varying human preferences. Direct Preference Optimization (DPO) [20] is one widely adopted method for this: it refines model responses by comparing two alternatives, a chosen and a rejected response, and optimizing the model to favor the preferred one. This preference-based learning paradigm has been shown to improve output quality in various language and vision-language tasks, suggesting its potential for enhancing RRG performance as well.

Given the high variability in clinical guidelines and reporting styles, we assume that even randomly sampling a rejected response from the same dataset could positively impact model performance after DPO. Herein, we refer to this approach as Random DPO (RDPO). We introduce it as a universal and model-agnostic training framework that requires only a dataset of ground-truth image-text pairs and a pretrained RRG model checkpoint, eliminating the need for reward models, human preference annotations, or any extra training data.

We empirically show how RDPO improves three publicly available models on the largest relevant datasets, such as MIMIC-CXR [10] and CheXpert Plus [3]. The gain it provides is consistent and is up to 5% in terms of clinically relevant metrics. We summarize our contributions as follows:

- We propose Random DPO, a framework for enhancing RRG models, without requiring additional labels, rewards, or preference annotations.
- We set a new state-of-the-art result by taking the strongest publicly available checkpoint MAIRA-2 [1] and applying our method on top of it.

Related Work. Over the years, classification and segmentation tasks have been effectively addressed by deep learning techniques, showing robust performance in identifying various pathologies [15]. However, the automatic generation of textual radiology reports remains largely unsolved, despite recent progress in assessing report quality [28]. Emerging vision-language models, such as MAIRA-2 [1] and CheXagent [4], demonstrate significant promise in bridging the gap between image analysis and report generation, yet they are still not sufficiently optimized to meet the rigorous standards required for widespread clinical adoption.

Early deep learning (DL)-based approaches used convolutional neural networks (CNNs) to generate radiology reports [7]. Next, recurrent neural network (RNN) methods, including a hierarchical CNN-RNN model described in [12], advanced chest X-ray report generation. Attention-enhanced RNN architectures further refined outputs by focusing on clinically relevant features [27]. Finally, large-scale vision-language models (VLMs), which integrate specialized image encoders and large language models (LLMs) via a trained encoder, show promising results [1,4].

2 Method

2.1 Background

From an objective standpoint, RRG models are typically optimized using standard cross-entropy loss for next-token prediction [5]. This naturally encourages the model to follow an auto-regressive decoding process, generating the ground-truth caption step by step [23]. However, in practice, this choice of objective function introduces unintended biases. The model may learn common text patterns that frequently co-occur and erroneously associate them with visual patterns that are not necessarily related to the text. These biases arise due to both the limitations of the training data (such as its quantity and diversity of sources), as well as domain shifts in clinical data distributions.

In general domain, Large Language Model training is not limited by supervised fine-tuning (SFT) on a narrow dataset. Instead, it is followed by the alignment step, which *aligns* model's responses with the human values. Among many alternatives, Direct Preference Optimization [20] is considered to be the most simple to reproduce. It allows one to get rid of an extra reward model and directly use preference dataset. In addition to input prompts, it consists of both chosen, or preferred, responses and rejected, or negative, responses. It does not imply that rejected response is an incorrect answer to the given prompt, but it means that according to some subjective criteria this specific option is less preferable.

In our task, fully identical cases are extremely rare—at most, they differ in minor details or wording. This assumption, supported by empirical evidence, motivates experiments where random samples are treated as rejected ones. Surprisingly, this approach leads to quality improvements across multiple metrics, regardless of the pretrained checkpoints used as a starting point or the test datasets evaluated.

2.2 Formal Description

We denote starting model checkpoint as π_{ref}, and optimized model, initialized from this checkpoint, as π_θ. At each training step, we sample a batch of image-text pairs of size N: $X = \{(x_i, y_i)\}_{i=1}^{N}$ from the training set. To construct pairs for preference optimization, for each pair (x_i, y_i) we sample a random y_{j_i} from $\{y_k\}_{k=1}^{N}$, different from y_i. We can rename y_i as y_i^w - chosen response, and y_{j_i} as y_i^l - rejected response, and get DPO loss:

$$\mathcal{L}(\theta) = -\mathbb{E}_{X \sim D} \left[\sum_{i=1}^{N} \log \sigma \left(\beta \log \frac{\pi_\theta(y_i^w \mid x_i)}{\pi_{ref}(y_i^w \mid x_i)} - \beta \log \frac{\pi_\theta(y_i^l \mid x_i)}{\pi_{ref}(y_i^l \mid x_i)} \right) \right]. \quad (1)$$

In our implementation, we do not need to prepare response pairs in advance. Instead, we sample pairs at the batch collation step without need of allocating extra space. Our approach does not depend on choice of pretrained model π, source of training data D, or alternative Preference Optimization algorithm. Thus, the framework can be modified according to one's needs. Importantly, the framework does not require any extra data preparation or annotation and simply leverages datasets already collected for SFT, image captioning, or fine-tuning.

3 Experiments

3.1 Datasets

In our experiments, we use four publicly available datasets that provide CXR images paired with radiological reports: MIMIC-CXR [10], CheXpert Plus [3], IU X-ray [5] and Interpret-CXR. These datasets were collected from diverse clinical settings, enabling comprehensive evaluation of the RRG methods.

MIMIC-CXR contains 377,110 images corresponding to 227,835 radiographic studies. These studies are sourced from the emergency department of the Beth Israel Deaconess Medical Center in Boston, MA, and each study is accompanied by a corresponding radiological report.

CheXpert Plus comprises 223,228 unique pairs of radiology reports and chest X-rays, derived from 187,711 studies for 64,725 patient, acquired in both inpatient and outpatient centers. The curation and release of CheXpert Plus were supported in part by the Medical Imaging and Data Resource Center (MIDRC).

IU X-Ray is a publicly available dataset containing 7,470 pairs of radiology reports and chest X-rays, collected by Indiana University.

Interpret-CXR is a composite dataset that integrates MIMIC-CXR, CheXpert Plus, PadChest [2], and BIMCV COVID-19 [24]. PadChest contributes 104,393 X-ray studies, interpreted and reported by radiologists at the Hospital San Juan in Spain between 2009 and 2017. BIMCV COVID-19 adds 46,727 chest X-ray image-text pairs from the Data Bank of Medical Images of the Valencian Community. Since Interpret-CXR does not have a publicly available test set, we use it exclusively for alignment.

MIMIC-CXR and CheXpert Plus are used in both RDPO alignment and evaluation, while IU X-Ray is left for evaluation purposes due to its small size. For all datasets, we use the officially published train-val-test splits, ensuring consistency and reproducibility of our experiments.

3.2 Metrics

Table 1. RDPO results on the MIMIC-CXR test set.

Method	Datasets	RDPO	BLEU	Bert-score	Semb-score	Radgraph-comb	RadCliQ-v1
Stanford	MIMIC-CXR	✗	0.141	0.360	0.381	0.155	0.779
Stanford	MIMIC-CXR	✓	0.140	0.360	0.409	0.159	**0.799***
Stanford	MIMIC+CheXpert	✗	0.140	0.369	0.396	0.168	0.807
Stanford	MIMIC+CheXpert	✓	0.142	0.367	0.416	0.169	**0.820***
Stanford	CheXpert Plus	✗	0.076	0.306	0.331	0.118	0.681
Stanford	CheXpert Plus	✓	0.074	0.306	0.336	0.121	**0.685***
CSIRO	MIMIC-CXR	✗	0.145	0.326	0.368	0.149	0.736
CSIRO	MIMIC-CXR	✓	0.146	0.326	0.383	0.152	**0.746***
MAIRA-2	private	✗	0.162	0.411	0.431	0.195	0.909
MAIRA-2	Interpret-CXR	✓	0.166	0.416	0.446	0.198	**0.932***

In this work, we use five metrics: BLEU [19], Bert-score [29], Semb-score [6], Radgraph-combined [9] and RadCliQ-v1 [28]. BLEU remains a widely accepted n-gram-based metric that measures lexical overlap to provide a rough assessment of linguistic similarity. Semb-score evaluates semantic alignment at the concept level by comparing CheXbert's embeddings, emphasizing the preservation of meaning within the generated text. Radgraph-combined integrates graph-based representations of radiological entities and their relationships, offering a holistic view of both factual correctness and contextual coherence. RadCliQ-v1 is a composite metric created to be correlated with experts' preferences. For the consistency with other metrics (*higher is better*), we report inverted RadCliQ-v1.

3.3 Experimental Design

The starting point for our experiments are model checkpoints, trained on image-captioning task in the SFT manner. To make experiments fair, for primary tests we chose models, trained only on public RRG datasets on RRG task only. This strategy allows us to use the same training data and eliminate negative effect of potential catastrophic forgetting.

Stanford AIMI's Models from Chexpert-Plus Release. [3] - a vision encoder-decoder model. Encoder is Swinv2 [14], decoder is BertDecoder [21] with 2 layers. We do experiments with three checkpoints, trained on combinations MIMIC-CXR and Chexpert-Plus.

Cvt2Distilgpt2 (CSIRO). [16] - a vision encoder-decoder model. There are two checkpoints publicly available - trained on MIMIC and IU-Xray, respectively. We omit the IU-Xray checkpoint due to the size of training set insufficient for DPO. Encoder was initialized from Convolutional Transformer [25], while decoder was initialized from DistilGPT-2 [22]

MAIRA-2. As a proof-of-concept, we tried to replicate our results on the strongest publicly available RRG model, MAIRA-2 [1], despite that it was partially trained on private data. It is a LLaVa-like model [13], trained on multiple datasets and tasks, demonstrates superior performance across multiple benchmarks. Vision encoder is DINOv2 [18] pretrained on mix of public and private radiographs. LLM decoder is Vicuna-7b-v1.5.

Training Hyperparameters. We conducted all fine-tuning experiments using LoRA [8] with $\alpha = 16$ and $r = 8$ and over q and v weights of attention layers. Learning rate was set to $1e-6$ with cosine scheduler, batch size to 64, and number of training epochs to 3. Sampling was not used during inference.

3.4 Results

Table 2. RDPO results on the CheXpert Plus test set.

Method	Datasets	RDPO	BLEU	Bert-score	Semb-score	Radgraph-comb	RadCliQ-v1
Stanford	MIMIC-CXR	✗	0.117	0.243	0.320	0.106	0.626
Stanford	MIMIC-CXR	✓	0.126	0.255	0.380	0.132	**0.673***
Stanford	MIMIC+CheXpert	✗	0.127	0.286	0.417	0.179	0.746
Stanford	MIMIC+CheXpert	✓	0.147	0.295	0.464	0.190	**0.790***
Stanford	CheXpert Plus	✗	0.122	0.279	0.385	0.160	0.711
Stanford	CheXpert Plus	✓	0.124	0.279	0.403	0.160	**0.720**
CSIRO	MIMIC-CXR	✗	0.101	0.193	0.254	0.099	0.569
CSIRO	MIMIC-CXR	✓	0.108	0.206	0.291	0.118	**0.598***
MAIRA-2	private	✗	0.113	0.285	0.327	0.136	0.674
MAIRA-2	Interpret-CXR	✓	0.127	0.295	0.362	0.152	**0.707***

We measured performance when trained on MIMIC-CXR, CheXpert Plus, and their combination. Since MAIRA-2 is originally trained on a large mix of private data, we fine-tuned it using Interpret-CXR collection [26]. Our results show a consistent improvement in the primary evaluation metric (inverted RadCliQ-v1) across these datasets; see Table 1 and 2, with * indicating a statistically significant improvement (P-value < 0.05) under one-sided Wilcoxon signed-rank test. This suggests that leveraging CXR reports with the proposed RDPO approach, as an additional step following pre-training and SFT, enables models to extract additional useful information from texts, leading to better performance.

However, the IU X-Ray dataset presents a distinct challenge. Due to its stylistic differences from MIMIC-CXR and CheXpert, performance there varies significantly. Specifically, when we observe substantial improvements on MIMIC-CXR or CheXpert, the performance on IU X-Ray often declines (Table 3). This suggests that domain shifts between datasets play a crucial role in model generalization. Incorporating IU X-Ray into training could potentially lead to a more consistent improvement across all three datasets. But the limited size of its training set (7k compared to 200k in the others) poses a major constraint.

3.5 Radiologist Assessment

In addition to numerical evaluation on validation subsets, we conducted a small-scale expert evaluation. A board certified radiologist with 15 years of experience reviewed 100 MIMIC-CXR samples processed by the original *stanford-mimic-chexpert* model and our RDPO-aligned model, assessing clinical quality and completeness of reported findings. The RDPO model was preferred in 33 cases, the baseline in 7, with the remaining cases resulting in ties.

Table 3. RDPO results on the IU X-Ray test set.

Method	Datasets	RDPO	BLEU	Bert-score	Semb-score	Radgraph-comb	RadCliQ-v1
Stanford	MIMIC-CXR	✗	0.208	0.433	0.592	0.173	**1.074**
Stanford	MIMIC-CXR	✓	0.212	0.433	0.578	0.181	1.070
Stanford	MIMIC+CheXpert	✗	0.207	0.466	0.587	0.212	**1.201**
Stanford	MIMIC+CheXpert	✓	0.212	0.456	0.592	0.203	1.164
Stanford	CheXpert Plus	✗	0.165	0.414	0.490	0.129	0.894
Stanford	CheXpert Plus	✓	0.174	0.419	0.509	0.138	**0.926***
CSIRO	MIMIC-CXR	✗	0.201	0.406	0.611	0.195	1.077
CSIRO	MIMIC-CXR	✓	0.212	0.417	0.620	0.234	**1.171***
MAIRA-2	private	✗	0.197	0.474	0.603	0.218	**1.255**
MAIRA-2	Interpret-CXR	✓	0.198	0.466	0.600	0.217	1.231

3.6 Examples

In Figs. 1 and 2 we provide qualitative examples of how RDPO improves clinical accuracy metrics. RDPO writes more details about pathologies present on the study instead of mimiquing standard cliches.

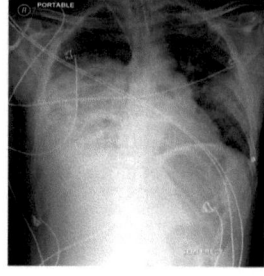

GT: As compared to the previous radiograph, the patient has received a new orogastric tube. The course of the tube is unremarkable, the tip of the tube projects over the middle parts of the stomach. There is no evidence of complications. The other monitoring and support devices are in unchanged position. *The massive bilateral lung abnormalities are constant in appearance.*

Baseline: as compared to the previous radiograph, the patient has received a nasogastric tube. the course of the tube is unremarkable, the tip of the tube projects over the middle parts of the stomach. the other monitoring and support devices are constant. *constant appearance of the lung parenchyma* and the cardiac silhouette.

RDPO: endotracheal tube terminates approximately 3.8 cm above the carina. an orogastric tube courses below the diaphragm, tip is not included in this examination. a right internal jugular catheter terminates in the region of the cavoatrial junction. *there is a large right pleural effusion with adjacent atelectasis. there is also a small to moderate right pleural effusion. there is also a small left pleural effusion.* there is no pneumothorax. the heart and mediastinum are within normal limits despite the projection.

Fig. 1. Test sample from MIMIC-CXR, generated by stanford-mimic-chexpert model. RDPO generates report with more clinical details.

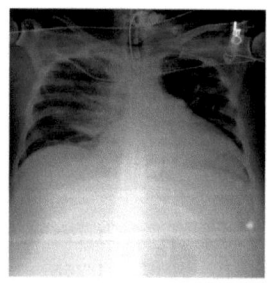

GT: Single AP portable chest radiograph is obtained. Tracheostomy tube is present. There is no pneumothorax or pleural effusion. *There is a hazy veil-like opacity in the right upper lung zone which may be consolidation, atelectasis or artifact.* Heart size appears enlarged; however, this may be technical due to AP view. Bony structures are intact.

Baseline: Single frontal view of the chest demonstrates an ET tube extending 3.5 cm above the carina. A left PICC tip is in the mid SVC. Lung volumes are low, accentuating bronchovascular crowding. There is no pneumothorax or large effusion. *There is no confluent consolidation to suggest pneumonia.* There is no pulmonary edema.

RDPO: Single frontal view of the chest demonstrates an ET tube extending 3.7 cm above the carina. An enteric tube extends inferiorly out of view into the stomach. The heart is enlarged, accentuated by low lung volume. Perihilar vascular markings are prominent, consistent with mild edema. *There is retrocardiac opacity, which could represent atelectasis or consolidation.* There is no pneumothorax or large effusion.

Fig. 2. Test sample from MIMIC-CXR, generated by MAIRA-2. RDPO generates a more clinically accurate report.

4 Conclusion

This work introduces RDPO—a novel alignment framework for VLMs in the context of RRG problem. In contrast to the existing alignment methods, e.g. DPO, our approach does not rely on additional annotation of human preferences.

Instead, RDPO relies just on SFT datasets comprising image-text pairs. In other words, RDPO allows to get more use out of these datasets than a regular SFT. Our experiments demonstrate that applying RDPO to several publicly available SoTA VLMs further increases their RRG quality. We emphasize that these results can be achieved without the cost of additional human annotation.

We believe that our framework can be further improved and scaled over larger data collections, leading to even more fascinating results. Alternative strategies for sampling a rejected response are also worth studying.

References

1. Bannur, S., et al.: MAIRA-2: grounded radiology report generation. arXiv preprint arXiv:2406.04449 (2024)
2. Bustos, A., Pertusa, A., Salinas, J.M., Iglesia-Vaya, M.: PadChest: a large chest X-ray image dataset with multi-label annotated reports. Med. Image Anal. **66**, 101797 (2020)
3. Chambon, P., et al.: CheXpert Plus: augmenting a large chest X-ray dataset with text radiology reports, patient demographics and additional image formats. arXiv preprint arXiv:2405.19538 (2024)
4. Chen, Z., et al.: CheXagent: towards a foundation model for chest X-ray interpretation. arXiv preprint arXiv:2401.12208 (2024)
5. Demner-Fushman, D., et al.: Preparing a collection of radiology examinations for distribution and retrieval. J. Am. Med. Inform. Assoc. **23**(2), 304–310 (2016)
6. Endo, M., Krishnan, R., Krishna, V., Ng, A.Y., Rajpurkar, P.: Retrieval-based chest X-ray report generation using a pre-trained contrastive language-image model. In: Machine Learning for Health, pp. 209–219. PMLR (2021)
7. Gale, W., Oakden-Rayner, L., Carneiro, G., Bradley, A.P., Palmer, L.J.: Producing radiologist-quality reports for interpretable artificial intelligence. arXiv preprint arXiv:1806.00340 (2018)
8. Hu, E.J., et al.: LoRA: low-rank adaptation of large language models. ICLR **1**(2), 3 (2022)
9. Jain, S., et al.: RadGraph: extracting clinical entities and relations from radiology reports. arXiv preprint arXiv:2106.14463 (2021)
10. Johnson, A.E., et al.: MIMIC-CXR, a de-identified publicly available database of chest radiographs with free-text reports. Sci. Data **6**(1), 317 (2019)
11. Ko, C.H., Chien, L.N., Chiu, Y.T., Hsu, H.H., Wong, H.F., Chan, W.P.: Demands for medical imaging and workforce size: a nationwide population-based study, 2000–2020. Eur. J. Radiol. **172**, 111330 (2024)
12. Liu, G., et al.: Clinically accurate chest X-ray report generation. In: Machine Learning for Healthcare Conference, pp. 249–269. PMLR (2019)
13. Liu, H., Li, C., Wu, Q., Lee, Y.J.: Visual instruction tuning. Adv. Neural. Inf. Process. Syst. **36**, 34892–34916 (2023)
14. Liu, Z., et al.: Swin Transformer V2: scaling up capacity and resolution. In: Proceedings of the IEEE/CVF Conference on Computer Vision and Pattern Recognition, pp. 12009–12019 (2022)
15. Meedeniya, D., Kumarasinghe, H., Kolonne, S., Fernando, C., Torre Díez, I., Marques, G.: Chest X-ray analysis empowered with deep learning: A systematic review. Appl. Soft Comput. **126**, 109319 (2022)

16. Nicolson, A., Dowling, J., Koopman, B.: Improving chest X-ray report generation by leveraging warm starting. Artif. Intell. Med. **144**, 102633 (2023)
17. Nicolson, A., Liu, J., Dowling, J., Nguyen, A., Koopman, B.: e-Health CSIRO at RRG24: entropy-augmented self-critical sequence training for radiology report generation. arXiv preprint arXiv:2408.03500 (2024)
18. Oquab, M., et al.: DINOv2: learning robust visual features without supervision. arXiv preprint arXiv:2304.07193 (2023)
19. Papineni, K., Roukos, S., Ward, T., Zhu, W.J.: BLEU: a method for automatic evaluation of machine translation. In: Proceedings of the 40th Annual Meeting of the Association for Computational Linguistics, pp. 311–318 (2002)
20. Rafailov, R., Sharma, A., Mitchell, E., Manning, C.D., Ermon, S., Finn, C.: Direct preference optimization: your language model is secretly a reward model. Adv. Neural. Inf. Process. Syst. **36**, 53728–53741 (2023)
21. Rothe, S., Narayan, S., Severyn, A.: Leveraging pre-trained checkpoints for sequence generation tasks. Trans. Assoc. Comput. Ling. **8**, 264–280 (2020)
22. Sanh, V., Debut, L., Chaumond, J., Wolf, T.: DistilBERT, a distilled version of BERT: smaller, faster, cheaper and lighter. In: NeurIPS EMC^2 Workshop (2019)
23. Selivanov, A., Rogov, O.Y., Chesakov, D., Shelmanov, A., Fedulova, I., Dylov, D.V.: Medical image captioning via generative pretrained transformers. Sci. Rep. **13**(1), 4171 (2023). https://doi.org/10.1038/s41598-023-31223-5, https://doi.org/10.1038/s41598-023-31223-5
24. Vayá, M.D.L.I., et al.: BIMCV COVID-19+: a large annotated dataset of RX and CT images from COVID-19 patients. arXiv preprint arXiv:2006.01174 (2020)
25. Wu, H., et al.: CVT: introducing convolutions to vision transformers. In: Proceedings of the IEEE/CVF International Conference on Computer Vision, pp. 22–31 (2021)
26. Xu, J., et al.: Overview of the first shared task on clinical text generation: RRG24 and "discharge me!". In: The 23rd Workshop on Biomedical Natural Language Processing and BioNLP Shared Tasks. Association for Computational Linguistics, Bangkok, Thailand (2024)
27. Xue, Y., et al.: Multimodal recurrent model with attention for automated radiology report generation. In: Frangi, A.F., Schnabel, J.A., Davatzikos, C., Alberola-López, C., Fichtinger, G. (eds.) MICCAI 2018. LNCS, vol. 11070, pp. 457–466. Springer, Cham (2018). https://doi.org/10.1007/978-3-030-00928-1_52
28. Yu, F., et al.: Evaluating progress in automatic chest X-ray radiology report generation. Patterns **4**(9) (2023)
29. Zhang, T., Kishore, V., Wu, F., Weinberger, K.Q., Artzi, Y.: BERTScore: evaluating text generation with BERT. In: International Conference on Learning Representations (2019)
30. Zhang, X., et al.: ReXrank: a public leaderboard for AI-powered radiology report generation. arXiv preprint arXiv:2411.15122 (2024)

Test-Time Adaptation of Medical Vision-Language Models

Fereshteh Shakeri[1,2](✉), Ghassen Baklouti[1,2], Julio Silva-Rodríguez[1], Maxime Zanella[3], Houda Bahig[2], Jose Dolz[1,2], and Ismail Ben Ayed[1,2]

[1] ETS Montreal, Montreal, Canada
fereshteh.shakeri.1@etsmtl.net
[2] CRCHUM, Montreal, Canada
[3] UCLouvain, Brussels, Belgium

Abstract. Integrating image and text data through multi-modal learning has gained significant attention in medical imaging research, building on its success in computer vision. While substantial progress has been made in developing medical foundation models and enabling their zero-shot transfer to downstream tasks, the potential of test-time adaptation remains largely underexplored in this domain. Inspired by recent advances in transductive learning and parameter-efficient fine-tuning, we investigate test-time adaptation of medical vision-language models. Our method leverages information-theoretic principles, maximizing the mutual information between the visual inputs and the text-based class representations, while minimizing a Kullback-Leibler divergence term penalizing deviation of the predictions from the zero-shot outputs. Building on this foundation, we introduce the first structured benchmark for test-time adaptation of medical vision-language models, exploring strategies tailored to the unique challenges of medical imaging. Our extensive experiments include two medical modalities, three specialized foundation models, six downstream tasks, and multiple state-of-the-art test-time adaptation methods, demonstrating significant performance improvements and establishing a new benchmark for this emerging field. The code is available at: https://github.com/FereshteShakeri/TTAMedVLMs.

Keywords: Medical Vision-Language Models · Test Time Adaptation · Transductive Inference · Parameter-Efficient Fine-tuning

1 Introduction

Deep neural networks have gained significant attention over the past decade in the medical image analysis community [19], driven by their transformative success in natural image recognition. Their advancements have been effectively applied to various medical imaging tasks, including tumor grading in gigapixel-stained histology images [30], diabetic retinopathy classification [7], and radiology image interpretation [13], among others. However, despite their potential,

---

F. Shakeri and G. Baklouti—Equal contributions.

© The Author(s), under exclusive license to Springer Nature Switzerland AG 2026
W.-K. Jeong et al. (Eds.): MedAGI 2025, LNCS 16112, pp. 181–191, 2026.
https://doi.org/10.1007/978-3-032-07845-2_18

the adoption of these models in real-world clinical settings remains limited. A major challenge lies in their dependence on large, annotated datasets, which are often difficult to obtain in medical domains [5]. Additionally, the presence of significant domain shifts—arising from inter-scanner variability, staining differences, and population heterogeneity—necessitates continuous adaptation for reliable performance across diverse clinical environments. A promising solution to address these challenges is transfer learning, where pre-trained models are leveraged to extract robust feature representations. While widely successful in computer vision, direct transfer from natural to medical images has not yielded the anticipated improvements [27], largely due to the fine-grained nature of medical imaging data and the unique complexities inherent to clinical datasets.

The field of transfer learning is undergoing a major transformation, driven by the emergence of foundation models that leverage large-scale pre-training on diverse datasets to enhance generalization across multiple tasks. Among these, vision-language models (VLMs) such as CLIP [26] and ALIGN [14] have demonstrated exceptional adaptability, utilizing contrastive learning to align visual and textual representations at scale. These models exhibit strong robustness to domain shifts [26], making them particularly effective for zero-shot learning and adaptable to low-shot settings or inference-time adjustments. However, despite their success in natural image domains, their direct application to medical imaging has been severely limited, primarily due to the lack of fine-grained medical expertise encoded within these models. Unlike general-purpose datasets, medical data requires specialized clinical knowledge, which current VLMs fail to capture, leading to suboptimal performance in real-world clinical settings.

To address this gap, significant efforts have been made to develop domain-specific medical VLMs, leveraging large open-access medical datasets across various specialties, including histology [11,12,21], ophthalmology [8,32,39], and radiology [35,36,45]. While these models provide a strong foundation, their adaptation to downstream medical tasks remains a challenge. Some studies have explored few-shot adaptation, either by training on a reduced percentage of the available datasets (e.g., 1% or 10% of samples in [12,45])—which still requires hundreds or thousands of annotated examples—or by establishing few-shot learning benchmarks [28] aligned with standard adaptation protocols. However, few-shot learning still requires labeled data, which may not always be available in real-world clinical scenarios. In contrast, test-time adaptation (TTA) offers a more practical and scalable alternative, allowing models to dynamically adapt to new domains without the need for labeled target data [34]. By adjusting model parameters [25] or refining predictions [42] based on incoming test samples, TTA can mitigate distributional shifts at inference time, making it particularly suitable for medical imaging applications, where domain variability is a persistent challenge. While TTA has gained increasing attention in computer vision, with recent and early explorations of adapting VLMs for natural image classification [25,42,44], its potential for medical imaging remains largely unexplored.

In this work, we introduce the first structured benchmark for test-time adaptation of medical VLMs, systematically evaluating adaptation techniques across multiple modalities and datasets. Inspired by recent advances in parameter-efficient fine-tuning (PEFT) and transductive learning, we investigate a novel language-aware test-time adaptation method. The approach maximizes the mutual information between the visual inputs and the text-based class representations, while preserving the model's zero-shot capabilities through a Kullback-Leibler divergence regularization. Our contributions could be summarized as follows:

○ We present the first structured test-time adaptation benchmark for medical vision-language models, including two medical imaging domains: histology, and ophthalmology.
○ We evaluate a range of test time adaptation strategies, including entropy minimization and transductive learning approaches, providing a comprehensive comparison of state-of-the-art TTA methods.
○ We investigate a novel language-aware adaptation framework that enhances test-time adaptation by maximizing the mutual information between the input images and the text-based class descriptions, while preserving the zero-shot knowledge.
○ We conduct extensive experiments on six downstream tasks using specialized foundation models, demonstrating that our language-aware TTA method consistently outperforms existing adaptation techniques.

2 Related Work

Medical Vision-Language Models. Building large-scale medical vision-language datasets is significantly more challenging than in natural image domains, where models like CLIP [26] and ALIGN [14] are trained on hundreds of millions of image-text pairs. In medical imaging, privacy regulations, specialized expertise, and limited annotations often constrain dataset availability. To address this, recent efforts have focused on developing vision-language models specifically for medical applications, spanning various domains such as radiology [35,36,45], ophthalmology [32,39], and computational pathology [11,12]. Notably, Quilt-1M [12], one of the largest vision-language histopathology datasets, aggregates one million image-text pairs from online sources, while ViLReF [39] is pretrained on over 450,000 retinal images with the corresponding diagnostic reports.

Test-Time Adaptation of Vision-only Models. Adapting models to new incoming data at test time has gained significant attention, with one of the initial works, TENT [34], which minimizes the entropy of predictions by updating the batch-norm statistics. Afterwards, LAME [4] highlighted the problem of model collapse when facing class imbalance, and proposes to only adapt the predictions, using Laplacian regularization. SAR [24] improves TENT with a sharpness-aware entropy loss, encouraging convergence to flatter loss regions and improving generalization. While these techniques primarily focused on natural-image tasks, they have also inspired research in medical imaging [2,10,16,20,

33,38]. While these advances in test-time adaptation can still bring significant improvements to medical vision-language models, tailored methods benefiting from the text modality are needed to fully leverage the adaptation capacity of these models.

Test-Time Adaptation for VLMs. Following the success of test-time adaptation in vision-only models and the remarkable zero-shot capabilities of CLIP, some approaches have emerged to adapt vision-language models at test time, but their application remains limited to natural images. One key distinction among these methods lies in the parameters they modify, ranging from input prompts [29] and intermediate layers, as in WATT [25], to adapters serving as memory banks [17] and training-free strategies [43]. A second distinction lies in how they process the incoming data. In this work, we focus on the setting where predictions must be made independently for each batch. This setup is particularly well-suited for transductive learning. For instance, TransCLIP [41,42] models each class with a multivariate Gaussian distribution, while enforcing consistency between the predicted labels and zero-shot predictions.

3 Methods

Zero-Shot Classification. At inference time, a vision language model performs zero shot classification by leveraging a simple yet effective text-image alignment strategy to generate final predictions. Given a set of k candidate classes, textual descriptions, a.k.a prompts, are created for each class. For instance, if one of the classes is *cancer*, the corresponding prompt might be [c_k = *a histopathology slide showing cancer cell.*]. These prompts are then processed through the text encoder θ_t, producing the corresponding text embeddings $t_k = \theta_t(c_k)$. Simultaneously, each test image x_i is passed through the vision encoder θ_v, yielding the visual embedding $f_i = \theta_v(x_i)$. The posterior probability of class k given input test image x_i is given by:

$$p_{i,k} = \frac{\exp(l_{i,k}/\tau)}{\sum_j^K \exp(l_{i,j}/\tau)} \qquad (1)$$

where τ is a temperature scaling parameter and $l_{i,k} = f_i^\top t_k$ is the logit score derived from the cosine similarity between the visual and text embeddings.

Proposed Objective for TTA. To further enhance the zero-shot predictions at inference time, we minimize a language-aware information maximization objective function tailored for the test time adaptation of VLMs, which we introduced recently in the context of natural-image few-shot tasks [1]:

$$-I(\theta_v, \theta_t) + T(\theta_v, \theta_t) \qquad (2)$$

In the following, we describe each of the two terms of the objective function above. $I(\theta_v, \theta_t)$ represents the mutual information between the vision inputs and the text-based class descriptions, considering them both as random variables, denoted as $\mathcal{X}_v$ and $\mathcal{X}_t$, respectively. It integrates conditional entropy

$H(\mathcal{X}_t|\mathcal{X}_v; \theta_v, \theta_t)$, which minimizes the uncertainty associated with the text-based prediction of each test sample, and marginal entropy $H(\mathcal{X}_t; \theta_v, \theta_t)$, which promotes a balanced class distribution, ensuring diversity in class predictions and preventing the model from favoring a limited number of classes. Formally, this reads:

$$I(\theta_v, \theta_t) = \lambda_{\text{ent}} H(\mathcal{X}_t) - \lambda_{\text{cond}} H(\mathcal{X}_t|\mathcal{X}_v)$$
$$= -\lambda_{\text{ent}} \sum_{k=1}^{K} \bar{p}_k \log \bar{p}_k + \lambda_{\text{cond}} \sum_{i \in B} \sum_{k=1}^{K} p_{i,k} \ln p_{i,k} \quad (3)$$

where λ_{ent} and λ_{cond} are two non-negative weighting factors that control the contribution of each term, and $\bar{p}_k = \frac{1}{|B|} \sum_{i \in B} p_{i,k}$ is the marginal probability of class k over the batch of test samples. B is the set of indices of the test samples in the batch. These entropy terms work together to promote confident and class-balanced predictions of the unlabeled test samples. In addition, regularization term $T(\theta_v, \theta_t)$ provides text-based supervision to preserve the model's zero-shot capabilities. It penalizes deviation of the model's predictions, $\mathbf{p}_i = (p_{i,k})_{1 \leq k \leq k}$, from the zero-shot ones, $\hat{\mathbf{y}}_i = (y_{i,k})_{1 \leq k \leq k}$. This is done by minimizing the following weighted Kullback-Leibler (KL) divergence:

$$T(\theta_t, \theta_v) = \lambda_{\text{txt}} \sum_{i \in B} \text{KL}(\mathbf{p}_i || \hat{\mathbf{y}}_i) = \sum_{i \in B} \mathbf{p}_i^\top \log \frac{\mathbf{p}_i}{\hat{\mathbf{y}}_i} \quad (4)$$

Also, similarly to the entropy terms, the text regularization term is weighted by a factor λ_{txt} to control its influence within our overall loss.

Fine-Tuning. To efficiently optimize the proposed loss, we adopt LoRA [9,40] as a fine-tuning strategy. This amounts to approximating the incremental updates ΔW of the original pre-trained weight matrix $W_0 \in \mathrm{R}^{m \times n}$ as the product of two low-rank matrices, A and B, based on the concept of "intrinsic rank" of a downstream task. Specifically, the low-rank weight modification is expressed as:

$$W = W_0 + \Delta W = W_0 + BA \quad (5)$$

where $A \in \mathrm{R}^{r \times n}$ and $B \in \mathrm{R}^{m \times r}$, with r representing the intrinsic rank, which is typically much smaller than m and n (i.e., $r << (m, n)$).

4 Experiments

4.1 Setup

Medical VLMs. Two predominant medical imaging modalities are included: histology, and ophthalmology. In each application, specialized VLMs that underwent modality-specific pre-training through textual supervision are leveraged. **Histology**: we employed Quilt-1M [12], which utilizes ViT-B/32 and ViT-B/16 vision encoder and a GPT-2 text encoder. **Ophthalmology**: we adopt ViLReF [39], a foundation model based on Chinese CLIP (CN-CLIP) [37], which has been pre-trained on several large-scale datasets containing approximately 200 million image-text pairs. ViLReF employs ViT-B/16 as its vision encoder and RoBERTa-wwm-ext-base-Chinese [6] as its text encoder.

Adaptation Tasks. We include a diverse set of tasks for TTA, designed to evaluate the transfer capabilities of medical VLMs. Each foundation model is applied to datasets within its respective medical domain. Note that all datasets are carefully selected to prevent test data leakage, ensuring that no samples used for evaluation were involved in the model's pre-training. In histology, the evaluation covers three different organs and cancer types, including colorectal adenocarcinoma (NCT-CRC) [18] , prostate cancer grading (SICAP-MIL) [31], and Lung cancer (LC25000-Lung) [3]. For ophthalmology, MESSIDOR [7] is focused on diabetic retinopathy (DR) grading. Also, FIVES [15] and ODIR200x3 [23], are incorporated to the benchmark, addressing inter-disease discrimination.

TTA Protocol and Evaluation. We conduct test-time adaptation following the standard practices established in vision literature [25]. A fixed batch size of 128 is employed throughout all experiments to ensure consistency and facilitate fair comparisons across different scenarios. Each batch is processed independently, and after adaptation and evaluation, the model is reset to prevent any information leakage between batches. Model performance is assessed using average class-wise accuracy (ACA), a widely used metric in medical imaging benchmarks [22]. All results are averaged over three random seeds across all experiments to ensure robustness and account for variability.

Implementation Details. The proposed method is deployed using a fixed LoRA rank of 2. For each input batch, 10 adaptation steps are carried out, updating the visual encoder's selected parameters. ADAM is the optimizer, using a learning rate of 10^{-3}. The marginal entropy weight is set to $\lambda_{\text{ent}} = 5$, the conditional entropy to $\lambda_{\text{cond}} = 1$ and the text regularization term is $\lambda_{\text{txt}} = 0.1$.

Baselines. A comprehensive evaluation of current TTA methods is carried out. Three different types of baselines are included. *i*) Zero-shot [26], i.e., no adaptation, where predictions are derived by computing the softmax cosine similarity between image and text embeddings. Text embeddings are generated using the specific prompts established in each medical VLMs. In particular, we employ prompt ensembles for Quilt-1M [12] and ViLReF [39], leveraging domain-specific textual representations to enhance alignment with visual features. *ii*) Blackbox TTA methods that rely uniquely on embedding representation. Transductive approaches LAME [4], and TransCLIP [42] are incorporated, whose hyperparameters are set as in the original publications. *iii*) Methods updating intermediate layers of vision or language encoders. We incorporate an adapted version of TENT [34] in which only the affine parameters in Layer Normalization (LN) layers are updated, applying entropy minimization directly to the model's logits. Notably, we perform 10 iterations of TENT adaptation per batch. We employ Adam optimizer with a fixed learning rate of 10^{-3}. Keeping this same learning configuration, we include SAR [24] and WATT [25]. We focus on WATT-p version, which optimizes the TTA loss separately for different models ($H = 4$ for histology, $H = 10$ for retina), each utilizing distinct textual templates.

4.2 Results

Performance Analysis. Table 1 provides a comprehensive comparison of our method against state-of-the-art TTA baselines in the histology and ophthalmology domains. *i)* **Comparison to black-box approaches**, i.e., LAME and TransCLIP. Our method achieves superior average performance across both modalities, significantly outperforming such baselines. TransCLIP, which operates without fine-tuning model parameters, performs well in cases where the number of classes is small and the zero-shot performance is already strong (e.g., three-class datasets such as LC-Lung and ODIR200x3). However, as the number of classes increases or when zero-shot performance is weaker, its effectiveness declines since it lacks the expressiveness of our proposed approach. This can be observed in SICAP-MIL with the Quilt-B32 model, where TransCLIP's performance drops by 10% compared to the zero-shot baseline. *ii)* **Comparison to parameter-efficient approaches**: TENT, SAR, and WATT. Our method outperforms TENT and SAR by large margins. For example, for histology tasks using Quilt-B/16 the average improvements are substantial: +23.8% and +19.3% respectively. Compared to WATT, our approach provides similar performance for histology tasks, but average improvements of +5.9% in ophthalmology. It is worth noting that while prior works show sometimes degrade its performance during TTA in at least one dataset, *the proposed method is the most robust, not suffering large degradations w.r.t. zero-shot predictions*.

Computational Efficiency. The results in Table 2 highlight the computational trade-offs between different adaptation strategies. LAME and TransCLIP exhibit the fastest evaluation times ($\sim$ 1.3 seconds), as they rely on lightweight optimization strategies without modifying model parameters. In contrast, TENT and SAR require slightly longer processing times since they require backpropagation through internal parameters in the vision encoder during adaptation. While achieving stronger adaptation performance, WATT incurs a significantly higher computational cost ($\sim$ 12.4 minutes), making it impractical for real-time applications. Our proposed method is computationally closer to TENT or SAR, maintaining a reasonable runtime, yet yielding outstanding results, even surpassing WATT. Overall, *our method demonstrates state-of-the-art performance while maintaining a balance between effectiveness and efficiency*, making it a compelling choice for test-time adaptation in medical vision-language models.

Robustness to Query Batch Size. Figure 1 illustrates the robustness of the different TTA methods when receiving data in smaller batches. Transductive strategies such as TransCLIP may suffer severe performance drops when accessing only small data batches since they cannot leverage the whole data distribution. In contrast, parameter-efficient strategies such as WATT and our proposed method are the more robust solution, providing satisfactory performance even

Table 1. Comparison of state-of-the-art TTA methods.

	Zero-Shot [26] ICLR'21	TENT [34] ICLR'21	SAR [24] ICCV'23	LAME [4] CVPR'22	TransCLIP [42] NeurIPS'24	WATT [25] NeurIPS'24	Ours
(a) Histology - Quilt-B/32 [12]							
NCT-CRC	52.62	60.33	60.12	60.57	58.56	59.96	**61.99**
LC-LUNG	80.57	79.37	80.73	85.68	**92.73**	87.52	89.02
SICAP-MIL	27.51	27.25	27.56	23.12	17.28	29.38	**29.45**
Avg.	53.57	55.65	56.14	56.46	56.19	58.95	**60.15**
(b) Histology - Quilt-B/16 [12]							
NCT-CRC	29.90	27.65	30.40	31.72	36.90	**49.67**	43.93
LC-LUNG	56.70	51.90	57.20	59.05	54.25	81.82	**85.98**
SICAP-MIL	32.78	28.40	33.60	28.54	48.26	43.90	**49.30**
Avg.	39.79	35.98	40.40	39.77	46.47	58.46	**59.74**
(c) Ophtalmology - ViLReF [39]							
MESSIDOR	42.66	39.31	36.29	45.82	28.89	26.88	**48.92**
FIVES	75.00	69.84	70.37	74.32	75.62	**75.99**	74.25
ODIR200x3	68.47	70.90	72.19	68.06	**87.43**	86.87	84.17
Avg.	62.04	60.02	59.62	62.73	63.98	63.25	**69.11**

when receiving query batches of 32 images. Considering the different modalities, our approach is the most stable solution for this challenging study.

Table 2. Computational workload. Runtime comparison of TTA methods for processing a batch of 128 samples (LC-LUNG dataset). All experiments were conducted on a single NVIDIA GeForce RTX 3090 (24GB) GPU.

	Black-box adaptation		Parameter-Efficient Fine-Tuning			
	LAME [4]	TransCLIP [42]	TENT [34]	SAR [24]	WATT [25]	Ours
Runtime (sec.)	~ 1.3	~ 1.3	~ 12.4	~ 13.5	~ 750.0	~ 15.1

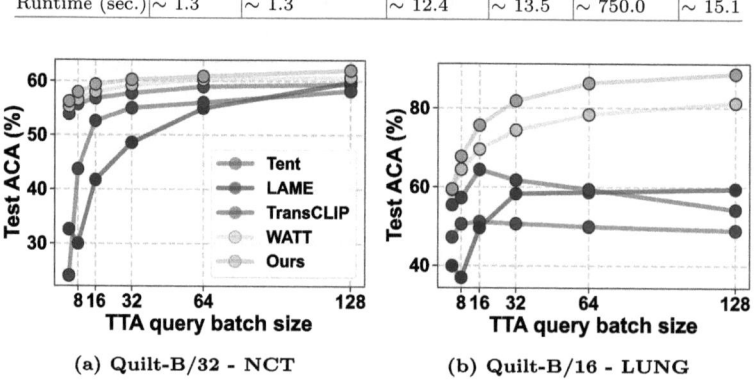

Fig. 1. Ablation Study. Effect of query batch size in TTA.

5 Conclusions

We introduced the first structured benchmark for test-time adaptation (TTA) of medical vision-language models (VLMs), systematically evaluating a range of adaptation strategies across multiple domains. In addition, we investigated a novel language-aware test-time adaptation framework that maximizes the mutual information between visual inputs and textual class representations, while preserving the model's zero-shot capabilities. Extensive experiments on histology and ophthalmology data showed that our method consistently outperforms existing adaptation techniques. Additionally, our computational analysis highlights that this does not come at the price of a high computational cost. Our findings underscore the importance of using both visual and textual modalities for more robust adaptation, paving the way for future research on test-time adaptation for medical VLMs.

Acknowledgments. This work was funded by the Natural Sciences and Engineering Research Council of Canada (NSERC) and Montreal University Hospital Research Center (CRCHUM). We also thank Calcul Quebec and Compute Canada.

Disclosure of Interests. The authors have no competing interests to declare that are relevant to the content of this article

References

1. Baklouti, G., Zanella, M., Ayed, I.B.: Language-aware information maximization for transductive few-shot clip (2025)
2. Bateson, M., Lombaert, H., Ben Ayed, I.: Test-time adaptation with shape moments for image segmentation. In: MICCAI, pp. 736–745 (2022)
3. Borkowski, A.A., et al.: Lung and colon cancer histopathological image dataset (lc25000). arXiv preprint arXiv:1912.12142 (2019)
4. Boudiaf, M., Mueller, R., Ben Ayed, I., Bertinetto, L.: Parameter-free online test-time adaptation. In: CVPR. pp. 8344–8353 (2022)
5. Chen, X., et al.: Recent advances and clinical applications of deep learning in medical image analysis. Med. Image Anal. **79** (2022)
6. Cui, Y., Che, W., Liu, T., Qin, B., Wang, S., Hu, G.: Revisiting pre-trained models for Chinese natural language processing. In: EMNLP, pp. 657–668 (2020)
7. Decencière, E., et al.: Feedback on a publicly distributed image database: the messidor database. Image Anal. Stereol. **33**, 231–234 (2014)
8. Du, J., et al.: Ret-CLIP: a retinal image foundation model pre-trained with clinical diagnostic reports. In: MICCAI, pp. 709–719 (2024)
9. Hu, E.J., et al.: LoRA: low-rank adaptation of large language models. In: ICLR (2021)
10. Hu, M., et al.: Fully test-time adaptation for image segmentation. In: MICCAI, pp. 251–260 (2021)
11. Huang, Z., et al.: A visual–language foundation model for pathology image analysis using medical twitter. Nat. Med. **29**, 1–10 (2023)
12. Ikezogwo, W., et al.: Quilt-1M: one million image-text pairs for histopathology. NeurIPS **36**, 37995–38017 (2023)

13. Irvin, J., et al.: ChexPert: a large chest radiograph dataset with uncertainty labels and expert comparison. In: AAAI, vol. 33, pp. 590–597 (2019)
14. Jia, C., et al.: Scaling up visual and vision-language representation learning with noisy text supervision. In: ICML, pp. 4904–4916 (2021)
15. Jin, K., et al.: FIVES: a fundus image dataset for artificial intelligence based vessel segmentation. Sci. Data **9**, 475 (08 2022)
16. Karani, N., et al.: Test-time adaptable neural networks for robust medical image segmentation. Med. Image Anal. **68**, 101907 (2021)
17. Karmanov, A., Guan, D., Lu, S., El Saddik, A., Xing, E.: Efficient test-time adaptation of vision-language models. In: CVPR, pp. 14162–14171 (2024)
18. Kather, J.N., Halama, N., Marx, A.: 100,000 histological images of human colorectal cancer and healthy tissue. Zenodo10 **5281** (2018)
19. Litjens, G., et al.: A survey on deep learning in medical image analysis. Med. Image Anal. **42**, 60–88 (2017)
20. Liu, Q., et al.: Single-domain generalization in medical image segmentation via test-time adaptation from shape dictionary. In: AAAI, vol. 36, pp. 1756–1764 (2022)
21. Lu, M.Y., et al.: Visual language pretrained multiple instance zero-shot transfer for histopathology images. In: CVPR, pp. 19764–19775 (2023)
22. Maier-Hein, L., et al.: Metrics reloaded: recommendations for image analysis validation. Nature Meth. **21** (2024)
23. NIHDS-PKU: Competition on Ocular Disease Intelligent Recognition (ODIR). https://odir2019.grand-challenge.org (2019)
24. Niu, S., et al.: Towards stable test-time adaptation in dynamic wild world. In: ICLR (2023)
25. Osowiechi, D., et al.: WATT: weight average test time adaptation of CLIP. NeurIPS **37**, 48015–48044 (2025)
26. Radford, A., et al.: Learning transferable visual models from natural language supervision. In: ICML, pp. 8748–8763 (2021)
27. Raghu, M., Zhang, C., Kleinberg, J., Bengio, S.: Transfusion: understanding transfer learning for medical imaging. In: NeurIPS, pp. 1–11 (2019)
28. Shakeri, F., et al.: Few-shot adaptation of medical vision-language models. In: MICCAI, pp. 553–563 (2024)
29. Shu, M., et al.: Test-time prompt tuning for zero-shot generalization in vision-language models. NeurIPS **35**, 14274–14289 (2022)
30. Silva-Rodríguez, J., et al.: Going deeper through the Gleason scoring scale: an automatic end-to-end system for histology prostate grading and cribriform pattern detection. Comput. Meth. Progr. Biomed. **195**, 105637 (2020)
31. Silva-Rodríguez, J., et al.: Proportion constrained weakly supervised histopathology image classification. Comput. Biol. Med. **147**, 105714 (2022)
32. Silva-Rodriguez, J., et al.: A foundation language-image model of the retina (FLAIR): encoding expert knowledge in text supervision. Med. Image Anal. **99**, 103357 (2025)
33. Valanarasu, J.M.J., et al.: On-the-fly test-time adaptation for medical image segmentation. In: MIDL, pp. 586–598. PMLR (2024)
34. Wang, D., Shelhamer, E., Liu, S., Olshausen, B., Darrell, T.: Tent: fully test-time adaptation by entropy minimization. In: ICLR (2021)
35. Wang, Z., et al.: MedCLIP: contrastive learning from unpaired medical images and text. In: EMNLP, pp. 1–12 (2022)
36. Wu, C., Zhang, X., Zhang, Y., Wang, Y., Xie, W.: MedKLIP: medical knowledge enhanced language-image pre-training for X-ray diagnosis. In: ICCV (2023)

37. Yang, A., et al.: Chinese CLIP: contrastive vision-language pretraining in Chinese. arXiv preprint arXiv:2211.01335 (2022)
38. Yang, H., et al.: DLTTA: dynamic learning rate for test-time adaptation on cross-domain medical images. IEEE TMI **41**(12), 3575–3586 (2022)
39. Yang, S., et al.: VilReF: an expert knowledge enabled vision-language retinal foundation model. IEEE TMI (2024)
40. Zanella, M., Ben Ayed, I.: Low-rank few-shot adaptation of vision-language models. In: CVPRw, pp. 1593–1603 (2024)
41. Zanella, M., et al.: Boosting vision-language models for histopathology classification: predict all at once. In: MedAGI MICCAIw, pp. 153–162 (2024)
42. Zanella, M., et al.: Boosting vision-language models with transduction. NeurIPS **37**, 62223–62256 (2024)
43. Zanella, M., et al.: On the test-time zero-shot generalization of vision-language models: do we really need prompt learning?. In: CVPR, pp. 23783–23793 (2024)
44. Zanella, M., et al.: Realistic test-time adaptation of vision-language models. In: CVPR, pp. 25103–25112 (2025)
45. Zhang, Y., Jiang, H., Miura, Y., Manning, C.D., Langlotz, C.P.: Contrastive learning of medical visual representations from paired images and text. In: MHLC (2022)

MaskedCLIP: Bridging the Masked and CLIP Space for Semi-supervised Medical Vision-Language Pre-training

Lei Zhu[✉], Jun Zhou, Rick Siow Mong Goh, and Yong Liu

Institute of High Performance Computing (IHPC), Agency for Science, Technology and Research (A*STAR), 1 Fusionopolis Way, #16-16 Connexis, 138632 Singapore, Republic of Singapore
zhu_lei@ihpc.a-star.edu.sg

Abstract. Foundation models have recently gained tremendous popularity in medical image analysis. State-of-the-art methods leverage either paired image-text data via vision-language pre-training or unpaired image data via self-supervised pre-training to learn foundation models with generalizable image features to boost downstream task performance. However, learning foundation models exclusively on either paired or unpaired image data limits their ability to learn richer and more comprehensive image features. In this paper, we investigate a novel task termed semi-supervised vision-language pre-training, aiming to fully harness the potential of both paired and unpaired image data for foundation model learning. To this end, we propose **MaskedCLIP**, a synergistic masked image modeling and contrastive language-image pre-training framework for semi-supervised vision-language pre-training. The key challenge in combining paired and unpaired image data for learning a foundation model lies in the incompatible feature spaces derived from these two types of data. To address this issue, we propose to connect the masked feature space with the CLIP feature space with a bridge transformer. In this way, the more semantic specific CLIP features can benefit from the more general masked features for semantic feature extraction. We further propose a masked knowledge distillation loss to distill semantic knowledge of original image features in CLIP feature space back to the predicted masked image features in masked feature space. With this mutually interactive design, our framework effectively leverages both paired and unpaired image data to learn more generalizable image features for downstream tasks. Extensive experiments on retinal image analysis demonstrate the effectiveness and data efficiency of our method.

Keywords: Foundation Model · Semi-Supervised Vision-Language Pre-training · Retinal Image Analysis

1 Introduction

Deep neural networks [28] have been a fundamental tool in medical image analysis, yet they often require a large amount of labeled training data to be effective and the models can sometimes be biased towards the semantic labels. Foundation models [3] provide a promising approach to alleviate these issues via pre-training deep neural networks on diverse and large volume of medical image data to learn generalizable image features to boost downstream task performance. State-of-the-art (SoTA) pre-training methods can be generally categorized into vision-language pre-training methods [34] and self-supervised pre-training methods [5,17,31,41]. Contrastive Language-Image Pre-training (CLIP) [34] is a leading vision-language pre-training method in general image analysis, where it leverages large-scale paired image-text data and contrastively aligns image features with text features in a shared feature space. Building on CLIP, numerous studies in medical domain have trained foundation models for different image modalities, including Chest X-ray [37], computed tomography (CT) [16], pathology [20], among others. More recently, FLAIR [35] proposes to encode expert knowledge to the text branch of CLIP for retinal image analysis, which boosts CLIP model performance. Self-supervised pre-training methods propose various pretext tasks to learn foundation models with unpaired image data. In general image analysis, contrastive learning based methods, such as SimCLR [8], SwAV [5], and MoCo-v3 [9], learn to maximize agreement between differently augmented samples in feature space. DINO [6] proposes self-distillation on multi-view images. Masked image modeling [17] pre-trains transformer to reconstruct masked image patches. iBot [40] performs self-distillation on masked image patches with an online tokenizer. DINOv2 [31] combines image-level [6] and patch-level [40] self-distillation with a novel data curation pipeline, which achieves state-of-the-art performance for various downstream tasks. More recently, RETFound [41] performs a systematic study to compare different self-supervised learning methods on retinal image analysis, where they found that generative based masked image modeling method [17] outperforms contrastive learning based ones [5,6,8,9].

While existing pre-training methods can train powerful foundation models to boost downstream task performance, they have focused exclusively on leveraging either paired or unpaired image data for learning foundation models, which limits their ability to learn richer and more comprehensive image features. In this paper, we investigate a novel task termed semi-supervised vision-language pre-training, aiming to fully harness the potential of both paired and unpaired image data for foundation model learning. To this end, we propose **MaskedCLIP**, a synergistic masked image modeling and contrastive language-image pre-training framework for semi-supervised vision-language pre-training. The key challenge in combining paired and unpaired image data for learning a foundation model lies in the incompatible feature spaces derived from these two types of data, where the CLIP feature space captures more semantic specific features, while the masked feature space retains more general image features. Thus, a naive approach to directly share the masked feature space with CLIP feature space suffers from the feature incompatibility issue and will result in poor performance. To address this issue, we propose to connect the masked feature space with the CLIP feature space with a bridge transformer. In this way, the more semantic specific CLIP features can benefit from the more general masked features for semantic feature

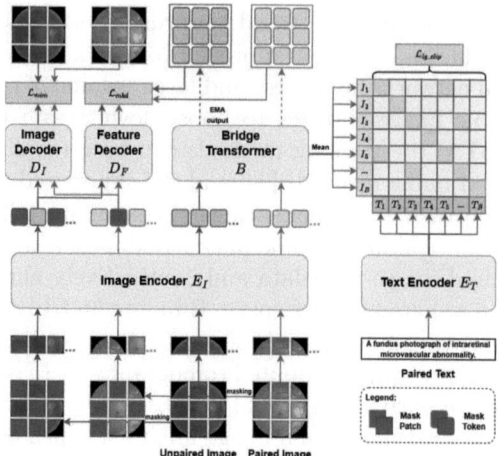

Fig. 1. Architecture and dataflow of our proposed MaskedCLIP framework. Our framework consists of five modules, namely an image encoder, a bridge transformer, a text encoder, an image decoder, and a feature decoder to process both image and text data for synergistic masked image modeling and contrastive language-image pre-training. We employ the bridge transformer to connect the masked and CLIP feature space to resolve the feature incompatibility issue and the feature decoder to predict masked image features for masked knowledge distillation.

extraction. We further propose a masked knowledge distillation loss to distill the semantic knowledge of original image features in CLIP feature space back to the predicted masked image features in the masked feature space. With this mutually interactive design, the masked features and CLIP features benefit from each other for feature representation learning, which enables our framework to learn more generalizable image features for downstream tasks.

In summary, we have made the following contributions in this paper: **(1).** We introduce a novel task termed semi-supervised vision-language pre-training for learning foundation models; **(2).** We propose MaskedCLIP, a principally designed framework for semi-supervised vision-language pre-training; **(3).** We conduct extensive experiments to evaluate the effectiveness of our method on retinal image analysis, where it significantly outperforms existing methods across seven downstream tasks and demonstrates exceptional label efficiency.

2 Methodology

In semi-supervised vision-language pre-training, we are give an assembly of paired image-text data with N^p data points. We represent the paired image-text data in a triplet format to accommodate optional categorical labels as $\mathbb{D}^p = \{(x_i^p, t_i^p, y_i^p)\}_{i=1}^{N^p}$, where t_i^p is the associated text description for x_i^p and y_i^p is the associated categorical label of x_i^p if available; otherwise y_i^p is defined as a unique identifier for the image-text pair. Additional, we are given an assembly of unpaired image data $\mathbb{D}^u = \{x_i^u\}_{i=1}^{N^u}$ with N^u data points. The goal is effectively leverage both the paired and unpaired image data to train a foundation model. Figure 1 presents an overview of our proposed MaskedCLIP framework.

2.1 Bridging the Masked and CLIP Space

Self-supervised pre-training and vision-language pre-training are two disparate learning paradigms that utilize either unpaired image data or paired image-text data for learning generalizable image feature representations. To effectively leverage both paired and unpaired image data for foundation model learning, one naive idea is to combine the two learning paradigms with a shared image encoder to learn a common image feature space. However, we note that there exists a natural semantic hierarchical structure across the feature spaces learned from the two types of data, where the vision-language pre-trained image features are more semantic specific due to language supervision, while self-supervised pre-trained image features are more general due to lack of supervision signal. Thus, naively sharing the image encoder from the two learning paradigms suffers from the feature incompatibility issue and will result in poor performance. To address this, we propose to bridge the two feature spaces while preserving their semantic hierarchical structure so that the more semantic specific vision-language image features can benefit from the more general self-supervised image features for semantic feature extraction. In our framework, we propose synergistic masked image modeling [17] and contrastive language-image pre-training [34] with a bridge transformer to bridge the masked and CLIP feature space.

Specifically, we utilize an image encoder E_I together with an image decoder D_I to construct the masked feature space for masked image modeling. We propose to combine paired image data together with the unpaired image data for the task to enhance data diversity. Following [17], the image encoder takes only the visible image patches as input and the image decoder takes concatenated latent features from visible image patches and learnable mask tokens with positional embedding as input to reconstruct the masked image patches. The masked image modeling loss is defined as follow:

$$\mathcal{L}_{mim} = \frac{1}{|\mathcal{B}^p| + |\mathcal{B}^u|} \sum_{x \in \mathcal{B}^p \cup \mathcal{B}^u} \frac{1}{|\mathcal{M}|} \sum_{i \in \mathcal{M}} ||x[i] - D_I([E_I(x_v); \boldsymbol{T}_I])[i]||^2, \quad (1)$$

where x_v denotes the visible image patches of x, $\boldsymbol{T}_I$ denotes the set of learnable mask tokens with positional embeddings for image pixel reconstruction, the operation $[\cdot;\cdot]$ concatenates two vectors into a single vector, $[\cdot]$ selects the indexed image patch from an image, $||\cdot||$ calculates the l_2-norm, $\mathcal{M}$ denotes the indexes of masked image patches, $\mathcal{B}^p$ and $\mathcal{B}^u$ denote batches of paired and unlabeled data sampled from $\mathbb{D}^p$ and $\mathbb{D}^u$ respectively.

Next, we introduce a bridge transformer B to connect the masked and CLIP feature space and a text encoder E_T to extract text features for contrastive language-image pre-training. Following [34], we employ a lightweight projection head at the end of both the bridge transformer and the text encoder to map the mean image and mean text features into a shared feature space. Vanilla contrastive loss [34] does not consider the categorical labels of different images, which can lead to images with the same categorical labels being erroneously pushed apart from the paired text of another image. Inspired from [38], we perform label-guided contrastive learning, where we ensure image-text pairs are pulled together if they share the same categorical label. We utilize the paired image-text data for pre-training, where we employ the image-to-text contrastive loss to align matched images to a given text and text-to-image contrastive loss to align matched texts to a given image. The loss functions are defined as follow:

$$\mathcal{L}_{i2t} = \frac{1}{|\mathcal{B}^p|} \sum_{(x,t)\in\mathcal{B}^p} \frac{1}{|\mathcal{P}(x)|} \sum_{(x',t')\in\mathcal{P}(x)} \log \frac{exp(\tau B(E_I(x))^T E_T(t'))}{\sum_{(x'',t'')\in\mathcal{B}^p} exp(\tau B(E_I(x))^T E_T(t''))},$$
(2)

$$\mathcal{L}_{t2i} = \frac{1}{|\mathcal{B}^p|} \sum_{(x,t)\in\mathcal{B}^p} \frac{1}{|\mathcal{P}(x)|} \sum_{(x',t')\in\mathcal{P}(x)} \log \frac{exp(\tau B(E_I(x'))^T E_T(t))}{\sum_{(x'',t'')\in\mathcal{B}^p} exp(\tau B(E_I(x''))^T E_T(t))},$$
(3)

where τ is a learnable scaling parameter and $\mathcal{P}(x) = \{(x',t')|(x',t') \in \mathcal{B}^p, y' = y\}$ is the set of image-text pairs with same categorical label as x within the batch.

The label-guided contrastive language-image pre-training loss is the combination of both image-to-text and text-to-image contrastive loss and is defined as follows:

$$\mathcal{L}_{lg\_clip} = \frac{1}{2}\mathcal{L}_{i2t} + \frac{1}{2}\mathcal{L}_{t2i}.$$
(4)

Discussion. While label-guided contrastive learning requires categorical labels as input, it reduces to the vanilla contrastive loss function when categorical labels are not available. Additional, for paired image-label data, we can apply simple prompt to convert categorical labels into text descriptions [34]. Thus, by incorporating label-guided contrastive learning into our framework, our framework works with both paired image-text and paired image-label data, which highlights the wide applicability of our approach in medical domain.

2.2 Masked Knowledge Distillation

We propose a masked knowledge distillation loss to further transfer semantic knowledge from original image features in CLIP feature space back to predicted masked image features in masked feature space. Such a loss function offers two key benefits: (1) It complements the pixel reconstruction loss in masked image modeling by guiding the model to extract semantic information from low-level image features for semantic feature reconstruction; (2) It enhances CLIP in global semantic information learning by extracting semantic information from local image patches. We leverage a feature decoder D_F to reconstructs the masked image features and propose to extract robust image features from original images in CLIP space as targets with the momentum encoder $\hat{M} = EMA(B \circ E_I)$, where $EMA(\cdot)$ calculates the exponential moving average of the encoder. We combine both the paired and unpaired image data for masked knowledge distillation. We normalize the masked image patch features and the target image patch features and minimize the cosine distance for all image patches for knowledge distillation. The masked knowledge distillation loss is defined as follows:

$$\mathcal{L}_{mfd} = \frac{1}{|\mathcal{B}^p| + |\mathcal{B}^u|} \sum_{x \in \mathcal{B}^p \cup \mathcal{B}^u} \frac{1}{|\mathcal{K}|} \sum_{i \in \mathcal{K}} -\langle \frac{D_F([E_I(x_v); \boldsymbol{T}_F])[i]}{||D_F([E_I(x_v); \boldsymbol{T}_F])[i]||}, \frac{\hat{M}(x)[i]}{||\hat{M}(x)[i]||} \rangle, \tag{5}$$

where $\boldsymbol{T}_F$ denotes the set of learnable mask tokens with positional embeddings for image feature reconstruction, $-\langle \cdot, \cdot \rangle$ calculates the cosine distance of two normalized vectors, and $\mathcal{K}$ denotes the indexes of all image patches.

Overall Objective. The overall objective of our MaskedCLIP framework is defined as follows:

$$\mathcal{L}_{maskedclip} = \mathcal{L}_{min} + \lambda_{lg\_clip}\mathcal{L}_{lg\_clip} + \lambda_{mfd}\mathcal{L}_{mfd}. \tag{6}$$

where λ_{lg_clip} and λ_{mfd} are two balancing weights. We empirically tune these two hyper-parameters based on their magnitudes and set them to 0.01.

3 Experimental Analysis

Pre-training Datasets. We assemble a pre-training dataset with 15 public datasets and 10 private datasets for retinal image analysis. In total, the assembled dataset comprises 348,481 color fundus images. The public datasets contain main retinal image analysis tasks in diabetic retinopathy grading [2,22,29], glaucoma detection [10,13,19,25,26,32,36], and some other disease diagnosis [4,21,27]. While most of the public datasets contain categorical labels, we also include two datasets that contain text descriptions: ODIR-5K [30] and STARE [18]. We build the paired image-text data with three public datasets, namely ODIR-5K [30], AIROGS [10], and EYEPACS [22], which contains 142,249 images in total. We follow [35] to encode expert knowledge to text descriptions when constructing image-text pair. We utilize the rest public datasets and all private datasets to build the unpaired image data.

Downstream Tasks and Comparison Methods. We evaluate the performance of our pre-trained foundation model on 7 public datasets across three retinal image analysis tasks: diabetic retinopathy grading (APTOS [23], IDRID [33], MESSIDOR-2 [11]), glaucoma detection (GF [1], ORIGA [39]), and multi-disease diagnosis (JSIEC [7], Retina [24]). We employ two commonly-used classification metrics, namely the area under receiver operating curve (ROC) and the area under precision-recall curve (PRC) to quantitatively evaluate the downstream task performance. We compare our method with a baseline Random method without pre-training; SoTA self-supervised pre-training methods MAE [17] and DINOv2 [31], where both methods are pre-trained on all paired and unpaired image data in our assembled dataset; SoTA vision-language pre-training method CLIP [34] which is pre-trained on the paired image data in our assembled dataset;

a baseline supervised pre-training method ImageNet21K [12], which is supervised pre-trained on about 14M labeled general images; SoTA foundation models in retinal image analysis, namely RETFound [41] pre-trained using MAE on about 0.9M color fundus images and FLAIR [35] pre-trained with encoded expert knowledge using CLIP on about 0.28M paired color fundus images.

Implementation Details. We implement the image encoder with ViT-large and the image decoder with ViT-small [14]. All input image is resized to 224×224. The patch size is set to 16×16. The masked ratio is set to 0.75 following [17]. The EMA parameter of the momentum encoder is set to 0.999. The bridge transformer and feature decoder are implemented using a vision transformer with 4 transformer blocks. The text encoder is implemented using BioClinical-BERT [15]. We train the model with AdamW optimizer for 200 epochs with a learning rate of 1.5e-4 and a warm-up period of 40 epochs. The model is trained on 4 NVIDIA A100 GPUs with a batch size of 720 (4 × 180 per GPU) for both paired and unpaired data. For downstream task fine-tuning, we initialize a ViT-large model with the pre-trained weights from our image encoder. We set the batch size to 16 and fine-tune the model with AdamW optimizer for 50 epochs with a learning rate of 5e-4 and a warm-up period of 10 epochs.

Table 1. Comparison with SoTA methods and foundation models on different downstream tasks. The best results are in **bold**, and the second-best results are underlined.

Labeled	Method	Paired Data Size	Unpaired Data Size	APTOS ROC	APTOS PRC	IDRID ROC	IDRID PRC	MESSIDOR-2 ROC	MESSIDOR-2 PRC	GF ROC	GF PRC	ORIGA ROC	ORIGA PRC	JSIEC ROC	JSIEC PRC	Retina ROC	Retina PRC	Avg ROC	Avg PRC
10%	Random	0	0	73.1	34.6	53.0	28.9	68.1	29.1	81.3	64.1	52.2	54.0	64.8	10.5	59.7	38.7	64.6	37.1
	CLIP [34]	0.14M	0	87.9	51.6	61.0	33.2	78.7	37.1	87.9	73.1	58.8	57.2	79.1	26.8	62.5	41.6	73.7	45.8
	DINOv2 [31]	0	0.34M	88.1	53.8	67.1	35.4	74.6	38.4	90.5	76.7	53.4	50.7	83.8	28.3	60.4	35.9	74.0	45.6
	MAE [17]	0	0.34M	91.9	58.8	72.7	40.7	79.2	44.3	86.8	73.5	53.2	54.2	85.6	38.2	67.7	48.7	76.7	51.2
	ImageNet21K [12]	14M	0	89.2	54.5	71.9	39.2	77.0	42.6	87.6	71.2	62.5	60.3	85.6	42.0	63.9	40.7	76.8	50.1
	FLAIR [35]	0.28M	0	90.2	54.3	63.9	33.6	76.2	41.7	84.8	67.1	59.3	57.6	83.1	31.6	68.3	47.6	75.1	47.7
	RETFound [41]	0	0.90M	91.4	61.9	66.8	38.1	78.1	41.8	88.6	74.1	65.4	62.9	87.1	41.5	68.5	45.9	78.0	52.3
	MaskedCLIP	0.14M	0.20M	**93.7**	**66.3**	**78.4**	**52.3**	**84.3**	**56.6**	88.0	71.9	**72.5**	**66.8**	85.2	35.8	**72.4**	**54.7**	**82.1**	**57.8**
100%	Random	0	0	83.7	45.8	56.6	30.5	69.9	30.4	87.1	72.0	52.4	52.9	81.9	27.9	61.8	40.8	70.5	42.9
	CLIP [34]	0.14M	0	92.0	64.6	75.9	44.8	83.9	52.8	92.4	82.5	61.6	58.7	97.6	76.5	82.8	66.5	83.7	63.8
	DINOv2 [31]	0	0.34M	94.2	70.7	77.1	46.8	85.2	59.5	**95.2**	**88.5**	64.0	58.1	99.3	89.6	81.9	66.3	85.3	68.5
	MAE [17]	0	0.34M	94.0	71.7	80.9	48.7	87.7	59.7	91.9	81.7	72.1	65.4	99.3	89.7	85.4	69.0	87.3	69.4
	ImageNet21K [12]	14M	0	94.3	70.3	78.0	47.6	86.3	60.9	91.5	80.7	64.8	60.4	**99.7**	**93.4**	80.5	60.6	85.0	67.7
	FLAIR [35]	0.28M	0	93.4	68.1	75.8	47.5	86.2	57.1	90.6	78.9	66.0	60.9	99.1	84.6	81.5	59.2	84.7	65.2
	RETFound [41]	0	0.90M	**95.0**	**74.4**	83.0	51.3	88.1	65.0	94.1	85.2	68.1	62.6	99.0	89.8	83.4	68.4	87.2	71.0
	MaskedCLIP	0.14M	0.20M	94.8	73.4	**83.1**	**56.3**	**88.8**	**68.5**	93.4	85.0	**72.8**	**66.2**	99.5	91.9	**90.4**	**79.5**	**89.0**	**74.4**

Table 2. Ablation study on different downstream tasks with 10% training data. The best results are in **bold**, and the second-best results are underlined.

Method	Bridge Transformer	$\mathcal{L}_{mfd}$	APTOS ROC	APTOS PRC	IDRID ROC	IDRID PRC	MESSIDOR-2 ROC	MESSIDOR-2 PRC	GF ROC	GF PRC	ORIGA ROC	ORIGA PRC	JSIEC ROC	JSIEC PRC	Retina ROC	Retina PRC	Avg ROC	Avg PRC
MAE+CLIP			91.5	59.3	72.6	39.2	82.1	43.2	87.8	71.8	66.7	60.9	81.3	26.9	65.7	42.6	78.2	49.2
+Bridge Transformer	✓		91.5	61.1	71.0	37.1	82.9	44.0	85.0	68.5	72.1	65.2	82.2	30.4	68.5	46.0	79.0	50.3
MaskedCLIP	✓	✓	**93.7**	**66.3**	**78.4**	**52.3**	**84.3**	**56.6**	**88.0**	**71.9**	**72.5**	**66.8**	**85.2**	**35.8**	**72.4**	**54.7**	**82.1**	**57.8**

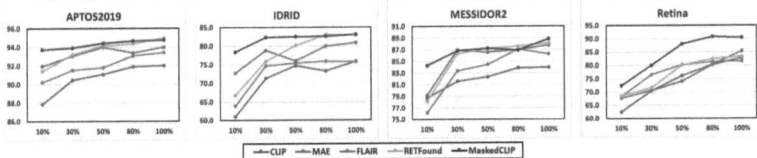

Fig. 2. Label efficiency analysis on exemplary downstream tasks. The X axis shows the training data proportion and the Y axis shows the ROC score.

Comparison with SoTA Methods. In Table 1, we compare our method with SoTA methods and foundation models under two learning scenarios, namely a label scarce setting with 10% training data and a label abundant setting with entire training data for fine-tuning across 7 downstream tasks. In both scenarios, our method consistently outperforms or matches existing methods and foundation models with significantly better average ROC and PRC scores. Specifically, our method outperforms SoTA self-supervised pre-training methods DINOv2 and MAE and vision-language pre-training method CLIP. Either DINOv2 and MAE or CLIP can only leverage unpaired or paired image data for pre-training, which limits their ability to learn richer and more comprehensive image features. In contrast, our method effectively integrates both paired and unpaired image data for foundation model learning, leading to significantly better performance. **The experiment results highlight that rather than focusing exclusively on either paired or unpaired image data, a unified approach that leverages all available image data is key to develop more powerful foundation models.** Furthermore, our method consistently outperforms or matches SoTA foundation models FLAIR and RETFound despite they are pre-trained with much larger paired or unpaired image datasets. The experiment results demonstrate that our method is more data efficient for foundation model learning than existing methods. We attribute this advantage to the integration of both paired and unpaired image data for pre-training and the novel mutually interactive design of our framework where the masked feature space is bridged to support CLIP feature space for semantic feature extraction and the CLIP feature space guides masked feature space for semantic feature learning through masked knowledge distillation. This mutually interactive design maximizes the utilization of both types of data. Finally, our method achieves the most improvement when compared to the second best in label scare setting, which highlights the label efficiency of our method for downstream tasks.

Ablation Study. In Table 2, we present an ablation study on different components of our method. As observed, directly combining MAE and CLIP by sharing their image encoder results in suboptimal performance. Introducing a bridge transformer to address the feature incompatibility issue between the two feature spaces significantly improves performance across multiple datasets. Finally, further incorporating masked knowledge distillation effectively enables the mutual interaction between the masked and CLIP feature spaces, which significantly enhances the performance across all downstream tasks and achieves the best results.

More Label Efficiency Analysis. In Fig. 2, we present a more detailed label efficiency analysis of our method against SoTA methods on APTOS2019, IDRID, MESSIDOR2, and Retina datasets. As shown, our method consistently outperforms or matches existing methods and foundation models in all training data proportions, highlighting its wide applicability in diverse learning scenarios.

4 Conclusion

In this paper, we introduce a novel task termed semi-supervised vision-language pre-training and propose MaskedCLIP, a principally designed framework to fully harness the potential of both paired and unpaired image data for foundation model learning. While existing studies have mostly focused on leveraging only paired or unpaired image data for learning foundation models, we advocate for a unified approach that integrates all available image data, either paired or unpaired to develop more powerful foundation models in medical domain. Our approach demonstrates promising results and we hope our work can inspire future studies to further explore this direction.

Acknowledgments. This work was supported by the Agency for Science, Technology, and Research (A*STAR) through its IEO Decentralised GAP Under Project I24D1AG085.

Disclosure of Interests. The authors have no competing interests to declare that are relevant to the content of this article.

References

1. Ahn, J.M., Kim, S., Ahn, K.S., Cho, S.H., Lee, K.B., Kim, U.S.: A deep learning model for the detection of both advanced and early glaucoma using fundus photography. PLoS ONE **13**(11), e0207982 (2018)
2. Benítez, V.E.C., et al.: Dataset from fundus images for the study of diabetic retinopathy. Data Brief **36** (2021)
3. Bommasani, R., et al.: On the opportunities and risks of foundation models. arXiv preprint arXiv:2108.07258 (2021)
4. Budai, A., Bock, R., Maier, A., Hornegger, J., Michelson, G.: Robust vessel segmentation in fundus images. Int. J. Biomed. Imag. (2013)
5. Caron, M., Misra, I., Mairal, J., Goyal, P., Bojanowski, P., Joulin, A.: Unsupervised learning of visual features by contrasting cluster assignments. Adv. Neural. Inf. Process. Syst. **33**, 9912–9924 (2020)
6. Caron, M., Touvron, H., Misra, I., Jégou, H., Mairal, J., Bojanowski, P., Joulin, A.: Emerging properties in self-supervised vision transformers. In: CVPR (2021)
7. Cen, L.P., et al.: Automatic detection of 39 fundus diseases and conditions in retinal photographs using deep neural networks. Nature commun. (2021)

8. Chen, T., Kornblith, S., Norouzi, M., Hinton, G.: A simple framework for contrastive learning of visual representations. In: ICML (2020)
9. Chen, X., Xie, S., He, K.: An empirical study of training self-supervised vision transformers. In: CVPR (2021)
10. De Vente, C., Vermeer, K.A., Jaccard, et al.: AIROGS: artificial intelligence for robust glaucoma screening challenge. IEEE Trans. Med. Imag. **43**(1), 542–557 (2023)
11. Decencière, E., et al.: Feedback on a publicly distributed image database: the messidor database. Image Analy. Stereol. (2014)
12. Deng, J., Dong, W., Socher, R., Li, L.J., Li, K., Fei-Fei, L.: Imagenet: a large-scale hierarchical image database. In: CVPR (2009)
13. Diaz-Pinto, A., Morales, S., Naranjo, V., Köhler, T., Mossi, J.M., Navea, A.: CNNs for automatic glaucoma assessment using fundus images: an extensive validation. Biomed. Eng. Online **18**, 1–19 (2019)
14. Dosovitskiy, A.: An image is worth 16 × 16 words: transformers for image recognition at scale. arXiv preprint arXiv:2010.11929 (2020)
15. Emily Alsentzer: (2019). https://huggingface.co/emilyalsentzer/Bio_ClinicalBERT
16. Hamamci, I.E., et al.: Developing generalist foundation models from a multimodal dataset for 3D computed tomography (2024)
17. He, K., Chen, X., Xie, S., Li, Y., Dollár, P., Girshick, R.: Masked autoencoders are scalable vision learners. In: CVPR (2022)
18. Hoover, A., Kouznetsova, V., Goldbaum, M.: Locating blood vessels in retinal images by piecewise threshold probing of a matched filter response. TMI **19**(3), (2000)
19. Huang, X., Kong, X., et al.: GRAPE: a multi-modal dataset of longitudinal follow-up visual field and fundus images for glaucoma management. Sci. Data **10**(1) (2023)
20. Ikezogwo, W., Seyfioglu, S., et al.: Quilt-1M: one million image-text pairs for histopathology. In: Neurips (2024)
21. Jin, K., et al.: FIVES: a fundus image dataset for artificial intelligence based vessel segmentation. Sci. Data **9**(1), 475 (2022)
22. Kaggle: (2015). https://www.kaggle.com/c/diabetic-retinopathy-detection
23. Kaggle: (2019). https://www.kaggle.com/c/aptos2019-blindness-detection
24. Kaggle: (2022). https://www.kaggle.com/datasets/jr2ngb/cataractdataset
25. Kumar, J.H., et al.: Chákṣu: a glaucoma specific fundus image database. Sci. Data **10**(1), 70 (2023)
26. Li, L., Xu, M., Wang, X., Jiang, L., Liu, H.: Attention based glaucoma detection: a large-scale database and CNN model. In: Proceedings of the IEEE/CVF Conference on Computer Vision and Pattern Recognition pp. 10571–10580 (2019)
27. Lin, L., et al.: The SUSTech-SYSU dataset for automated exudate detection and diabetic retinopathy grading. Sci. Data **7**(1), 409 (2020)
28. Litjens, G., Kooi, T., Bejnordi, B.E., et al.: A survey on deep learning in medical image analysis. Med. Image Anal. **42**, 60–88 (2017)
29. Liu, R., Wang, X., Wu, Q., Dai, L., Fang, X., et al.: DeepDRID: diabetic retinopathy—grading and image quality estimation challenge. Patterns **3**, 100512 (2022)
30. ODIR 2019 Grand Challenge: (2019). https://odir2019.grand-challenge.org/
31. Oquab, M., Darcet, T., et al.: DINOv2: learning robust visual features without supervision. arXiv preprint arXiv:2304.07193 (2023)

32. Orlando, J.I., Fu, H., Breda, J.B., Van Keer, K., et al.: REFUGE Challenge: a unified framework for evaluating automated methods for glaucoma assessment from fundus photographs. Med. Image Anal. **59**, 101570 (2020)
33. Porwal, P., et al.: IDRID: diabetic retinopathy-segmentation and grading challenge. Med. Image Anal. **59**, 101561 (2020)
34. Radford, A., et al.: Learning transferable visual models from natural language supervision. In: ICML (2021)
35. Silva-Rodriguez, J., Chakor, H., Kobbi, R., Dolz, J., Ayed, I.B.: A foundation language-image model of the retina (flair): encoding expert knowledge in text supervision. Med. Image Anal. **99**, 103357 (2025)
36. Sivaswamy, J., Krishnadas, S., Joshi, G.D., Jain, M., Tabish, A.U.S.: Drishti-GS: retinal image dataset for optic nerve head (ONH) segmentation. In: ISBI (2014)
37. Wang, Z., Wu, Z., Agarwal, D., Sun, J.: MedCLIP: contrastive learning from unpaired medical images and text. arXiv preprint arXiv:2210.10163 (2022)
38. Yang, J., Li, C., Zhang, P., Xiao, B., Liu, C., Yuan, L., Gao, J.: Unified contrastive learning in image-text-label space. In: Proceedings of the IEEE/CVF Conference on Computer Vision and Pattern Recognition, pp. 19163–19173 (2022)
39. Zhang, Z., Yin, F.S., et al.: ORIGA-light: an online retinal fundus image database for glaucoma analysis and research. In: EMBC (2010)
40. Zhou, J., Wei, C., Wang, H., Shen, W., Xie, C., Yuille, A., Kong, T.: IBOT: image BERT pre-training with online tokenizer. arXiv preprint arXiv:2111.07832 (2021)
41. Zhou, Y., et al.: A foundation model for generalizable disease detection from retinal images. Nature **622**, 156–163 (2023)

Author Index

A
Ayed, Ismail Ben 181

B
B. Gotway, Michael 1
Bahig, Houda 181
Bai, Jin 140
Baklouti, Ghassen 181
Belyaev, Mikhail 171
Bian, Jiang 87
Bobrin, Maksim 171
Boushehri, Ali 54
Brattoli, Biagio 44
Brehmer, Alexander 87

C
Cao, Xiaohuan 76
Carloni, Gianluca 44
Chen, Aokun 87
Chen, Zhu 12
Cheng, Joanna 140
Choe, Suhyun 98
Choi, Jaeha 98
Cluceru, Julia 54
Cohen, Yaniv 54
Contreras, Amanda Butler 87

D
Dada, Amin 87
Ding, Xiaowei 1
Dolz, Jose 181
Dylov, Dmitry 171

E
Edgü, Mert 12

F
Fan, Chenxiao 150
Fang, Jin 140

G
Ghamizi, Salah 109
Goh, Rick Siow Mong 192
Goncharov, Mikhail 171
Gotway, Michael B. 159
Gros, Eric 54

H
Hager, Gregory 140
Ho Ahn, Chang 44
Hu, Haoxi 65
Hu, Jingliang 65
Huang, Weijian 23

I
Ishii, Masaru 140

J
Jacob, Athira J. 120
Jin, Er 12
Jung, Sumin 34

K
Kanli, Georgia 109
Keum, Seongho 44
Keunen, Olivier 109
Keyl, Julius 87
Kikuchi, Yusuke 54
Kleesiek, Jens 87
Koraş, Osman Alperen 87

L
Lee, Jin Won 98
Lee, Seohyun 98
Lee, Taebum 44
Li, Chenjia 150
Li, Chunlei 65
Li, Yvonna 54
Liang, Jianming 1, 159
Liang, Yong 23

Liu, Jiarun 23
Liu, Yong 192
Luo, Haozhe 1

M
Ma, DongAo 159
Mangulabnan, Jan Emily 140
McLeod, Matthew 54
Mou, Lichao 65

N
Nie, Jianlong 76

O
Oseledets, Ivan 171

P
Pang, Jiaxuan 1, 159
Park, Jongchan 44
Peng, Cheng 87
Pereira, Sergio 44
Perquin, Magali 109

R
Rabah, Chaima Ben 130
Ren, Genqiang 76
Rueckert, Daniel 120

S
Samokhin, Valentin 171
Seenivasan, Lalithkumar 140
Seibold, Constantin 87
Serag, Ahmed 130
Shakeri, Fereshteh 181
Sharma, Puneet 120
Shen, Dinggang 76
Shen, Yiqing 150
Shi, Yilei 65
Shirokikh, Boris 171
Silva-Rodríguez, Julio 181
Smith, Kaleb E. 87

Soberanis-Mukul, Roger D. 140
Song, Dongning 23
Stegmaier, Johannes 12

T
Tae Kwak, Jin 34
Taher, Mohammad Reza Hosseinzadeh 1
Tan, Minhui 76
Tan, Tao 23
Taylor, Russell H. 140

U
Umerenkov, Dmitriy 171
Unberath, Mathias 140, 150
Usdin, Maxime 54

V
Vedula, S. Swaroop 140

W
Wang, Shanshan 23
Wu, Qingxia 76
Wu, Yonghui 87

X
Xiang Zhu, Xiao 65
Xue, Zhong 76

Y
Yang, Hao 23
Yang, Hyun 34
Yang, Qi 54

Z
Zanella, Maxime 181
Zhang, Boyang 76
Zhou, Hong-Yu 23
Zhou, Jun 192
Zhou, Ziyu 1, 159
Zhu, Lei 192

MIX
Papier aus verantwortungsvollen Quellen
Paper from responsible sources
FSC® C105338

If you have any concerns about our products,
you can contact us on
ProductSafety@springernature.com

In case Publisher is established outside the EU,
the EU authorized representative is:
**Springer Nature Customer Service Center GmbH
Europaplatz 3, 69115 Heidelberg, Germany**

Printed by Libri Plureos GmbH
in Hamburg, Germany